THE POLITICAL ECONOMY OF INTERNATIONAL AIR SAFETY

The Political Economy of International Air Safety

Design for Disaster?

Vicki L. Golich
Assistant Professor of Political Science
Pennsylvania State University

St. Martin's Press New York

Scholarly and Reference Division,
St. Martin's Press, Inc., 175 Fifth Avenue, New York, N.Y. 10010

First published in the United States of America in 1989

Printed in Great Britain

ISBN 0–312–02825–3

Library of Congress Cataloging-in-Publication Data
Golich, Vicki Lynne.
The political economy of international air safety
: design for disaster? / Vicki L. Golich
p. cm.
Bibliography: p.
Includes index.
ISBN 0–312–02825–3
1. Aeronautics—Accidents. 2. Aeronautics—Safety measures.
I. Title.
HE784.G65 1990
363.1′247—dc19

For Nate and Scott who will inherit our perceptions about and regimes for managing international relations

Contents

List of Tables

List of Figures

Preface and Acknowledgements

This book is concerned fundamentally with learning. How state elites learn new ways to perceive changed world and domestic environments and new ways to respond. How corporate elites learn new ways to perceive the global market structure and new ways to survive. How scholars learn new ways to perceive the relationship between structure and process and new ways to link them.

Hopefully among the lessons learned here, one will stand out – the need to distinguish between collective and individual rationality. Too often, as many scholars referenced within have articulated, acting according to individual rationality leads to collective irrationality. Collective irrationality ultimately hurts the individual.

Ours is one world. Whatever management scheme we dream up, it will serve us best if it facilitates compromise, co-operation and conflict resolution within a variable sum game framework which allows for mutual gains for all participants.

There are many to whom I am indebted for their advice and counsel as this project evolved. Throughout graduate school and then as a Post Doctoral Fellow at the University of Southern California's School of International Relations Research Institute I received valuable guidance and assistance from Jonathan Aronson, Gunnar Nielsson, William Tow and Alex Hybel. Each helped me gain focus with respect to the story I wanted to tell.

Alan Cafruny, Harvey Feigenbaum, Ernst Haas, Jeffrey Hart, Peter Katzenstein, Robert Keohane, James Lebovic, Stephen Krasner and Susan Strange each contributed specific suggestions or criticisms to parts of this work at different stages of completion.

Jerome Lederer, virtually the father of aviation safety engineering, taught me a lot about the politics and economics of aviation safety regulation, and how to differentiate between the ideal and the possible.

Not everyone mentioned will agree with the argument and conclusions made here. None should be held responsible for any errors of judgement or argument; but each receives my deepest thanks.

Two friends and colleagues deserve special mention because of their care of my psyche as well as my intellect: George Mitchell, my research assistant at USC, and Eileen Bresnahan, my teaching assistant and typist of infinite hours at the University of Colorado, Denver.

Barbara Ayala deserves tremendous thanks for patiently typing the multiple versions of this book until I was satisfied with every bit of minutiae.

Finally, my children – Nate and Scott – have earned my deepest gratitude and admiration for having learned at very young ages – this all began when Nate was 2 years old and Scott was not yet conceived! – to respect my request for a 'few hours of peace and quiet in the morning'.

State College, Pennsylvania VICKI L. GOLICH

1 Introduction: Power, Process and Crisis

> The central theme of international relations is not evil but tragedy. States often share a common interest, but the structure of the situation prevents them from bringing about the mutually desired situation. (Jervis, 1983: 20)

1974

On 3 March 1974, a Turkish Airline (THY-Turk Hava Yollari) DC10 initiated its Istanbul–Paris–London flight with about 200 passengers on board. Most were vacationers; 50 were going to Paris. Normally, few new travellers boarded THY for its Paris–London leg; the Paris stop was mostly a maintenance-necessitated 'pit-stop'. This time was different. A recent strike had grounded British European Airlines and wreaked havoc with flights all over Europe. Travellers attempting to return to London were forced into all kinds of re-routings, many through Paris' Orly International Airport. Once there, a great rush was on for any flight with a London destination. Consequently, THY Flight 981 started its takeoff roll with only 11 empty seats.

The DC10 lifted off Runway 08 just after 12:30 p.m. All indications were for a smooth and comfortable flight across the Channel. Suddenly, about nine minutes into the flight, at 11 500 feet, a loud explosion punctured the routine atmosphere. An improperly secured rear cargo door was blown out by the almost five tons of pressurised air pushing against the aircraft's sleek frame. The rapid decompression caused by the gaping hole of four-square-feet collapsed the cabin floor, which severed all the hydraulic control cables.

The flight deck crew never had any hope of recovery. The voice recorder reveals the experienced and capable pilots properly executed every manœuvre in their attempt to control the aircraft's violent movements. Only 77 seconds after the loss of the cargo door, THY Flight 981 slammed into the Ermenonville Forest at 497 miles per hour: 346 people from 21 different countries died. Most were young, under the age of 40; several entire families were wiped out.

1979

25 May 1979 was a beautifully clear, typically busy day at Chicago's O'Hare International Airport. In the early afternoon, thirteen American Airlines crew members readied their DC10 for a regularly scheduled flight to Los Angeles. Slowly, 258 passengers, many on their way to a writer's conference, filed on board. Maintenance personnel finished their routine tasks and monitored the aircraft's engine start, push-back, and initial taxi; no one noticed anything unusual. About 3 p.m. the crew was cleared for takeoff from Runway 32R, and the Captain acknowledged, 'American 191 under way.'

The takeoff roll was normal as the three engines surged toward their 40 000 pounds of thrust. Just before rotation, the point of actual lift-off, pieces of the left engine pylon structure fell to the ground. At the moment of rotation, the entire engine-pylon unit ripped loose from the wing. Powered by residual fuel in its lines, the turbofan shot ahead, arched up and over the wing, then crashed on the runway.[1] As it crossed over the wing, it destroyed hydraulic lines causing the left wing slats to retract, though the right wing slats were still extended in takeoff position. This asymmetrical slat situation destroyed the aircraft's slow lift capability. The engine-pylon trajectory also wrecked the electrical cables which otherwise would have activated the stall warning system.

Aware only that they had lost power in one engine, the crew initiated emergency procedures appropriate for that situation. Tragically, the crew could not know they had physically lost the engine and its pylon, nor could they be alerted to the aircraft's slat asymmetry and impending stall. This predicament required a different set of emergency manœuvres.

Flight 191 continued its climb to about 325 feet above the ground. Suddenly it rolled uncontrollably to the left and its nose pitched down below the horizon. About a minute after starting its takeoff roll, the American Airlines DC10 crashed into an open field just short of a trailer park and a few thousand feet beyond the end of the runway. All 271 persons on board were killed instantly; two people on the ground also died, and two more sustained second- and third-degree burns. The aircraft was demolished during the impact, explosion, and subsequent ground fire; wreckage was scattered everywhere. In addition, an old aircraft hangar, several automobiles, and a mobile home were destroyed.

AVIATION DISASTER AND INTERNATIONAL RELATIONS

Commercial airline crashes have catastrophic results, including hundreds of deaths and the losses of millions of dollars in aircraft and ground damages and lawsuits. They also shake the confidence of a world increasingly dependent on civil aviation as a crucial link in the transportation, trade, and communications chains circling the globe.

Virtually everyone has a personal interest in aviation safety; we all share a morbid fascination with the spectacular nature of aircraft disasters. This is a critical reason why I decided to examine international commercial aviation safety. My intent is to join in the scholarly effort to refine the international regimes framework into a parsimonious general theory of international relations. An inherently intriguing topic may capture our attention and force acknowledgment of actors and processes previously ignored.

A second reason for this focus is that international commercial aviation is intrinsically valuable to the world's political economy. When operating reliably and efficiently, it facilitates the trade regime which is vital to state survival. It is considered a high technology industry whose research and development efforts are perceived to be crucial components of economic growth. Trade of commercial aviation products generates more revenues than any other manufacturing effort. Air transport and tourism represent 'the largest single trade item after petroleum' (Jönsson, 1981: 275).

Third, state elites consider civil aviation to be a strategic industry because of its relationship to political, military, and economic security (see Chapter 3). Sectoral analyses,[2] particularly of strategic industries, illuminate key bargains and allow scholars to discover 'the variety of outcomes emerging' from them (Strange and Tooze, 1981: 12, 214). They highlight tensions which result from the ' "internationalisation" of national economic destinies coupled with a renewed demand by all kinds of social groups for participation in the processes which control their life chances' (*Ibid.*: 219). Sectoral analyses address the question of who gets what (Lasswell, 1965), which is as important to the study of international relations as to politics.

Fourth, international commercial aviation management is one of many issue areas in today's world political economy requiring state level co-operation to avoid destructive zero sum game outcomes.

The challenge of international air law will be to find the

> rules and the structures to facilitate . . . co-operation and to enable all mankind, in large nations and in small, to benefit from international aviation. (Salacuse, 1980: 843)

There is a tension between global and state level goals intrinsic to many of these issues. This analysis reveals the nature of this tension and indicates which incentive structures may impel or compel states to co-operate in the resolution of existing or potential conflict, and in the achievement of common goals when competitive, unilateral action might generate more immediate and tangible, if not enduring, benefits.

Finally, civil aviation is one of the world's few industries whose need for international management was recognised early. This is probably because Europe dominated the industry in its infancy and its geography made air transport international in scope almost from the very beginning.

I have chosen the specific issue of safety in commercial aviation for three reasons. First, it is an intrinsically fascinating issue in which we all have a personal interest.

More importantly, the need for safety in all aspects of aviation is fundamental to the industry's ability to provide the reliable and efficient transportation network required in today's interdependent world. Statistics such as those offered in Tables 1.1 and 1.2 ostensibly show that in spite of the fiery and spectacular nature of aircraft accidents, air transport is the safest existing means of travel.[3] The fact remains that our global skies could be safer than they are today.[4]

Third, the issue of safety in international commercial aviation embodies the divergent state and global level goals dilemma discussed above. These goals may not be as mutually exclusive as they are often perceived. It may be possible to achieve jointly satisfactory outcomes. The requisite bargaining process will be typical of many international relations negotiations. Safe, reliable, and efficient air transportation should be a common goal. Everyone – states, corporations, and the consuming public – will gain if it is; all will lose if it is not. Still, the strategic nature of the industry to individual states provides an incentive to act unilaterally and competitively in the struggle to create or maintain a healthy domestic aviation industry.

The 1974 and 1979 DC10 crashes described at the beginning of this chapter represent in dramatic fashion the ultimate failure of an aviation safety regime: each was preventable given the existing levels of technology and regime-prescribed rules and procedures. Each

Table 1.1 Safety in international commercial air transport by type of service, 1973–84[a]

Aircraft accidents involving passenger fatalities: Scheduled services

	Aircraft accidents	*Passenger fatalities*	*Passenger fatalities per 100 million passenger–km*
1973	36	862	0.17
1974	29	1 299	0.24
1975	20	443	0.08
1976	20[b]	734	0.12
1977	24	516	0.07
1978	25	755	0.09
1979	31	878	0.10
1980	21	812	0.09
1981	18	350	0.04
1982	25	748	0.08
1983	20[c]	809	0.08
1984	15[b]	223	0.02

[a] Excludes Union of Soviet Socialist Republics.
[b] Includes a mid-air collision shown as one accident.
[c] Includes one collision on ground counted as one accident.

Source: *ICAO Statistical Yearbook – 1984* (1985) p. 11.

helps identify the process and participants involved in the creation and maintenance of reasonable safety in international commercial air transport. Each indicates how politics and economics can interfere with the resolution of seemingly technological issues. Together they constitute 'disciplined configurative' case studies (Eckstein, 1975: 79–137; see also Smelser, 1973; Lijphart, 1971).

They also invite comparison. The 1974 crash was caused by a structural failure in the airframe and was linked to the manufacturer. The 1979 crash was caused by a failure in the maintenance process and was linked to the airline.

Aviation safety is dependent on many factors, including such intangibles as human competency and ability to cope with psychological and physiological stress. More tangible are technological factors. An enigma is that many available solutions to specific safety problems have not been integrated into the regime. Even more puzzling, existing safety regulations are often not rigorously

Table 1.2 Transportation accidents and resulting deaths and injuries by type of transport in the US, 1970–85

Year and Casualty	*Total (1 000)*	*Motor vehicles[1] (1 000)*	*Rail roads[2]*	*Air Carriers*				*General aviation[6]*	*Water borne[7]*
				Total	*Air lines[3]*	*Commuter air carriers[4]*	*On demand air carriers[5]*		
Accidents:									
1970	16 013	16 000	8 095	(NA)	(NA)	(NA)	(NA)	(NA)	(NA)
1975	16 524	16 500	8 041	237	37	48	152	3 995	3 310
1980	17 925	17 900	8 451	228	19	38	171	3 590	4 624
1981	18 020	18 000	5 781	214	26	31	157	3 500	3 503
1982	18 118	18 100	4 589	178	19	27	132	3 231	3 174
1983	18 319	18 300	3 906	184	25	19	140	3 060	4 704
1984	18 817	18 800	3 900	185	18	21	146	3 008	3 275
1985	(NA)	(NA)	3 426	194	24	17	153	2 748	2 259
Deaths:									
1970	54.8	52.6	785	(NA)	(NA)	(NA)	(NA)	(NA)	(NA)
1975	48.2	44.5	575	221	124	28	69	1 252	243
1980	54.5	51.1	584	143	1	37	105	1 239	206
1981	52.5	49.3	556	132	4	34	94	1 282	154
1982	47.2	43.9	512	320	234	14	72	1 182	223
1983	45.7	42.6	498	89	15	12	62	1 081	289
1984	44.6	44.3	598	102	4	46	52	1 115	113
1985	44.1	43.8	454	638	526	35	77	944	69
Injuries:									
1970	2 024	2 000	21 327	(NA)	(NA)	(NA)	(NA)	(NA)	(NA)
1975	1 858	1 800	54 306	109	71	6	32	728	97
1980	2 066	2 000	62 246	74	17	14	43	675	176
1981	1 956	1 900	53 003	82	21	24	37	597	141
1982	1 741	1 700	37 638	98	31	28	39	620	271
1983	1 636	1 600	32 196	49	8	12	29	566	209
1984	1 739	1 700	35 660	67	9	23	35	595	134
1985	1 735	1 700	31 617	73	10	12	51	497	57

NA Not available

[1] Data on deaths are from US National Highway Traffic Safety Administration and are based on 30-day definition.

[2] Train accidents cover only those accidents which result in damages to railroad property exceeding amounts specified by the US Federal Railroad Administration. The reporting threshold was raised from $750 to $1750 in 1975; to $2900 in 1979; to $3700 in 1981; to $4500 in 1983; and to $4900 in 1985. Deaths exclude fatalities in railroad-highway grade crossing accidents.

[3] Includes scheduled and non-scheduled (charter) air carriers operating under 14 CFR 121, 125, and 127; only serious injuries are included.

[4] All scheduled services operating under 14 CFR 135; only serious injuries are included.

[5] All non-scheduled services operating under 14 CFR 135; only serious injuries are included.

[6] 1975–84 data exclude commuter and on-demand air taxis operating under 14 CFR 135.

[7] Covers accidents involving commercial vessels which must be reported to US Coast Guard if there is property damage exceeding $1500; material damage affecting the sea-worthiness or efficiency of a vessel; stranding or grounding; loss of life; or injury causing a person's incapacity for more than three days.

Source: US Dept. of Transportation, Transportation Systems Center, Cambridge, MA *Transportation Safety Information Report* annual.

enforced, resulting in catastrophes. Why? Answers to this question will shed light on regime creation, maintenance, and change, and help refine this framework of analysis.

Throughout this inquiry I will focus on political and corporate actors and their agents of policy implementation. The political universe is limited to the United States (US) and the European Community (EC) and its member-states because of the preponderant roles each has played, historically and presently, in the global air transport regime.[5] Various EC states took an early lead in the development of aviation and its safety. Today they represent the most effective challenge to the 40 years of US industry dominance. Table 1.3 shows that the US and Europe are the only producers of commercial class, large transport aircraft in the Western World. Tables 1.4 and 1.5 demonstrate that, except for Japan, their airlines dominate the provision of global air transport services.

Table 1.3 Aircraft[a] orders and deliveries of commercial air carriers, 1984

Aircraft[b]	*Delivered*			*Ordered*[c]	*Remaining*[e]
	before 1984	*during 1984*	*Total 12/31/84*	*during 1984*	*to be delivered 12/31/84*
US Aircraft	5 170	187	5 357	283	557
European Aircraft	667	74	751	77	153
All Aircraft[d]	5 847	261	6 108	360	710

[a] Turbo-jet aircraft of 9 000 kg MTOW and over; 'kg MTOW' means 'Kilogram Maximum Take Off Weight'.
[b] Owing to lack of information the number of aircraft currently being manufactured in the USSR are not included in the table.
[c] Reported options are not included in the number of aircraft ordered.
[d] These figures exclude the cumulative totals of aircraft models that are no longer in production at 31 December 1983.
[e] The numbers in this column take into account cancellations during the year.

Source: *ICAO Statistical Yearbook* – 1982 (1983) pp. 34–47.

Aircraft manufacturers and airlines are the corporate actors examined here. Their products, airframes and air transport service are subject to a certification process which is supposed to ensure their safety and quality. A mutual dependency exists between the two actors: airlines need aircraft and manufacturers need consumers. What affects one will eventually affect the other. The industry dynamics of each differ in some ways, but for each there are powerful incentives to avoid regulation when possible. Since each is an integral part of the same strategic industry, there are also powerful incentives for governments to yield to industry pressures for regulation relaxation.

Aside from the fact that airframe manufacturers produce a tangible item and airlines produce an intangible transport service, the most important difference between the two is the number of producers. There are very few commercial class aircraft manufacturers globally: Boeing, McDonnell Douglas, and, until recently, Lockheed in the United States, and Airbus Industries and British Aerospace in Europe. There are many airlines globally: approximately 76 in the United States and at least one 'flag carrier' and often one or two more domestic carriers in virtually every other country in the world *(ICAO Statistical Yearbook, 1982*, 1983: 60–1; US Department of Transportation, 1987: 165–7). In previous studies, the number of producers was singled out as a significant variable, helping the scholar to explain and predict producer responses to political-economic events (Strange, 1979). Chapter 3 provides evidence that this variable is useful in analysing the responses of producers in the aviation industry as well. While important, this difference is not as significant with regard to safety issues as it may at first appear, because each industry segment is so dependent on the other for either to enjoy commercial success.

Finally, the agents of policy implementation include US and European state organisations responsible for creating and maintaining air transport safety, such as the US Federal Aviation Administration (FAA), the EC Joint Airworthiness Requirements Consortium (JARs), and the Dutch Civil Aviation Authority (RLD), etc. US agencies have long dominated international commercial aviation and virtually controlled its safety regime because nearly every state in the West subscribed to US Federal Aviation Regulations (FARS). Nevertheless, the co-operative efforts initiated within the EC in their production and certification of the Concorde and the Airbus are impressive and may represent a future behaviour model.

In the struggle to discern and articulate general theories of international relations, scholars seek parsimony and elegance.

Table 1.4 Total ton-kilometres performed for 1982 scheduled service[a]

Country	*Ton-kilometres performed*[b]	
	International	*International & domestic*
United States	94 569	456 105
European Community		
Belgium	6 244	6 244
Denmark[c]	2 875	3 605[e]
Federal Republic of Germany	23 121	25 371
France	32 060	43 509
Greece	4 169	5 431
Ireland	2 567	2 628
Italy	12 638[e]	17 096[e]
Luxembourg	101[e]	101[e]
Netherlands	18 840	18 871[e]
United Kingdom	57 261	60 251[e]
Others[d]		
Australia	17 268	28 219[e]
Brazil	7 391	19 458
Canada	16 037	35 953[e]
Israel	5 201	5 368[e]
Japan	30 959	62 717
Republic of Korea	13 474	14 281
Singapore	20 627	20 627
Spain	12 945	18 417
Switzerland	13 092	13 323

[a] Passengers freight and mail.
[b] Millions
[c] Denmark is a member of the consortium airline Scandinavian Airlines System (SAS); thus these figures include an appointment of the international traffic performed by SAS according to the formula used by the ICAO – 2/7 to Denmark 2/7 to Norway and 3/7 to Sweden. These figures also include Denmark's other airlines which reported data to ICAO; see Chapter 1, Note 2.
[d] Representative states of ICAO regions or significant amounts carried.
[e] Provisional or estimated data.

Source: *ICAO Statistical Yearbook – 1982* (1983) pp. 34–47.

Table 1.5 Total passenger-kilometres performed for 1982 scheduled service[a]

Country	*Ton-kilometres performed*[a]	
	International	*International & domestic*
United States	82 987	408 997
European Community		
Belgium	5 277	5 277
Denmark[b]	2 527	3 193
Federal Republic of Germany	19 595	21 625
France	27 438	37 846
Greece	3 767	4 924
Ireland	2 297	2 343
Italy	11 074[d]	15 143[d]
Luxembourg	92[d]	92[d]
Netherlands	16 253	16 282[d]
United Kingdom	50 892	53 645[d]
Others[c]		
Australia	15 355	25 266[d]
Brazil	6 368	17 229
Canada	14 224	32 140[d]
Israel	4 495	4 648[d]
Japan	26 491	55 731
Republic of Korea	11 364	12 101
Singapore	18 161	18 161
Spain	11 506	16 457
Switzerland	11 567	11 773
USSR	10 252	172 206

[a] Millions

[b] Denmark is a member of the consortium airline Scandinavian Airlines System (SAS); thus these figures include an appointment of the international traffic performed by SAS according to the formula used by the ICAO – 2/7 to Denmark 2/7 to Norway and 3/7 to Sweden. These figures also include Denmark's other airlines which reported data to ICAO; see Chapter 1, Note 2.

[c] Representative states of ICAO regions or significant amounts carried.

[d] Provisional or estimated data.

Source: *ICAO Statistical Yearbook – 1982* (1983) pp. 34–47.

While these are important goals in theory building, the various frameworks often lack enough detail to have much explanatory or forecasting value.

This research effort suggests that the international regimes framework as conceptualised by Robert Keohane and Joseph Nye (1977) and subsequently refined by many[6] is a useful starting place for understanding relations among actors in many important issue areas. This analysis incorporates some of these refinements; in particular, the following assertions are made.

First, the distribution of power capabilities helps determine which states participate in the creation of a regime governing any issue area. In contrast to the original Keohane and Nye contention, it is argued here that power capabilities can be fungible across issue areas. While US military strength may not be a useful tool for extracting economic trade and monetary concessions from Japan, there can be an indirect relationship: the US could threaten to retract its security umbrella from the Far East which would force Japan to invest more in defence and reduce what is available for commercial investment. Civil aviation is replete with the use of such *quid pro quos* – trade-offs in seemingly unrelated areas for gains in aviation – such as, a military base for expanded commercial landing rights. It may be that the propensity for issue area linkages is an important aspect of complex interdependence.

Subsequently, power capabilities help determine the degree to which states can successfully implement or impose policies on a global level. Very early in the evolution of civil aviation, an effort to create an international air law code similar to Grotius' conception of the freedom of the seas was effectively undermined by unilateral action by the United Kingdom which asserted airspace sovereignty instead. Similarly, in 1944, when 52 nations convened in Chicago to formalise international air transport rules and regulations, the United States, with its preference for an air transport system based on property rights, overrode the contending preference for treating it as an international public utility.

The second major assertion follows John Ikenberry's argument that states must adjust to both internal and external stimuli, and that they do so not as unitary actors, but rather as elites who must respond to domestic citizen demands in order to stay in office and authoritatively allocate values (1986: 53–77).[7] The history of civil aviation demonstrates that states respond to different constraints at different times. The following case studies shed some light on when and why states sometimes adjust to external structural constraints and at other times to internal, domestic ones.

A third assertion is that market conditions affect actor behaviour globally. The international regimes framework is particularly applicable because, as Chris Mason has argued, regimes are 'intermediate factors between the power structure of the market and the political and economic bargaining that takes place within it' (1979: 8). Therefore, the less 'magistracy' (Kindleberger, 1978) there is, the more vulnerable groups within states are to the structure of the global market. Both state and non-state actors react to perceived changes in market conditions. These responses may constrain the behaviour of other regime participants. For example, the case studies demonstrate how US policy choices, by supporting DC10 sales, resulted in the market failure of the L1011. Subsequently, Lockheed withdrew from commercial class aircraft production. Likewise, the creation of the Airbus consortium was a response to perceived changes in market conditions. Its success has required state level co-operation in an arena where nationalistic competition historically prevailed.

Fourth, international political economy scholars need to focus more attention on the process[8] of actor bargaining because of the insights to be gained regarding how and what states learn[9] about effective behaviour.[10] A focus on process will also help identify those changes in the international system or within states which trigger adjustment strategies. Some suggest that surplus capacity in a strategic industrial sector may be the catalyst (Strange, 1976; Strange and Tooze, 1981; Cowhey and Long, 1981). Other possibilities include advancing technology, a shift in ideas or knowledge, complacency, or other domestic factors.[11]

Most likely a convergence of several factors is necessary. Thus a final question to be addressed is, 'why do regimes change?' Since a number of studies indicate that they are created by dominant states in their own best interests, it has been an enduring enigma that they change. A number of possible explanations have been proffered: hegemonic decline, surplus capacity. These have been derived primarily from realist and economic analytical frameworks. An attempt is made to link, if not integrate, them more effectively.

Chapter 2 offers a history of civil aviation necessary to the understanding of the present regime. Here we discover how air transport emerged as an industry with political, military and economic strategic value to states. We also learn how the guiding norms, principles, rules and decision-making procedures of the regime were conceived and developed. A fascinating interplay of power capabilities helps explain why 'secondary states' deferred to US leadership in the creation and early maintenance of the regime (Keohane, 1984: 39).

Chapter 3 provides background information on the dynamics of the civil aviation industry. Several persistent political and economic pressures to short-circuit safety are identified and examined. The strategic nature of the industry is further explored making it clear why states respond the way they do to industry and citizen demands regarding appropriate state–industry relationships. This also reveals how states can change behaviours that have worked successfully for decades in an international environment because of domestic demands for such changes. Understanding the value and dynamics of the industry will shed light on the interactions between 'the system-level international economic structures and national level policy' (Lake, 1983: 539–40).

The case studies are presented in Chapter 4. These demonstrate how catalytic events can trigger state adjustment strategies. They reveal even more about the intricate relationship among the sundry factors affecting state and non-state actor behaviour.

Chapter 5 summarises and re-conceptualises the theoretical framework put forward by this study.

2 The Evolving Safety Regime

> Once policies have been adopted, they are pursued until a new crisis demonstrates that they are no longer feasible. States become locked in by the impact of prior choices on their domestic political structures. (Krasner, 1976: 341)

The evolution of the international commercial aviation regime spans seven decades. As the industry grew from a sporadic mail service to today's highly sophisticated means of transportation, the regime evolved from a series of separate national management schemes to one global in scope. The regime's development has been marked by a tension between state and global level needs and goals. States have always supported and encouraged the industry because of its strategic value; however the need for global coordination intrinsic to transportation systems has been just as constant.[1] The regime has vacillated between the unilateral assertion of state authority and multinational co-operation.

The safety regime is embodied in the aviation regime and can be seen in three phases: the first saw progress made by individual nations as each struggled to establish a domestic aviation industry.

The second began with the creation of international organisations, whose purposes included the promotion of standardised safety regulations. The attempt to manage aviation multilaterally was aborted by irreconcilable differences between the United States' (US) and European views regarding appropriate government – industry relationships. In the US, the prevailing philosophy was 'regulatory enterprise', which holds that state regulation should ensure the private operation and competition of airlines in an unfettered market. Europeans were guided by a 'participatory enterprise' concept which contends commerce is an instrument of national policy and requires government involvement (see Jönsson, 1981: 281). Global level management was finally achieved through a complex series of over 1500 bilateral agreements modelled after Bermuda I, the first Bilateral Air Transport Agreement negotiated between the United States and the United Kingdom (UK).

The third phase was ushered in by the renegotiated bilateral agreement between the US and the UK. Known as Bermuda II, it was a response to several factors. State complacency about regime management between 1946 and 1974 meant few policy changes had accompanied the radical advances in technology and knowledge that were causing structural shifts in the governing market conditions. Skyrocketing research, development and production costs in the manufacturing sector were aggravated by increased competition and excess capacity in the service sector. Adjustments by non-state actors placed new constraints on states which were beginning to adjust to structural changes in several other issue areas. This third period, still in transition, has been marked by states seeking to accommodate the need for increasingly close multilateral co-operation while keeping control over the authoritative allocation of values back home.

Regime participants are acutely aware of how important the current reorganising of the aviation regime is:

> Regimes are not only a means by which states can control outcomes *abroad* by influencing the behaviour of other states but also a means by which states can increase their power *at home* by means of agreements with other states about how economic activities are to be regulated, and by whom. (Mason, 1979: 4; emphases are his)

The following glimpse into aviation's past and present reveals who wrote the rules and why they were accepted. Such information provides important clues into the motivations for and likely outcomes of intra-regime negotiations and helps the researcher anticipate when significant changes may occur.

PHASE I: NATIONAL AIR SAFETY SOUGHT

Man's first successful flights were in lighter-than-air vehicles such as hot air balloons and helium-filled dirigibles, and prompted the first discussions regarding the establishment of an international air law code. At a 1902 meeting of the Institute of International Law in Brussels, the French suggested that Hugo Grotius' 1604 freedom of the seas principle be applied to the skies. No consensus could be reached and the idea of an international law of air space was dropped until after the era of heavier-than-air flight began with Orville Wright's first flight on 17 December 1903.

Initially, most thought the airplane had value only as a sport or military vehicle. Then, on 25 July 1909, the Frenchman Louis Bleriot made the first international flight across the English Channel, and a few began to toy with the idea of air transport. Within months, Germany's Count Zeppelin began the first passenger service.

At a 1910 international law conference, the French, with German support, again proposed a freedom-of-the-air doctrine. The British delegation, led by an admiral, favoured sovereign air space for military reasons (Cooper, 1947: 18–21; 1952: 127–43). The British acted quickly and decisively to 'safeguard their air space' by empowering the state 'to regulate the entry of foreign aircraft and to proscribe zones over which foreign aircraft were not allowed to fly' (Jönsson, 1981: 277). The European continent followed suit, and by 1914, most air space was closed.

Although the First World War temporarily halted the direct development of civil aviation, it encouraged military air transport. Both sectors benefitted from advances in aircraft technology because their needs coincided: reliable, safe, and speedy delivery of mail, goods, and people. Civilian aviation gains were by-products of enormous wartime government expenditures, when cost was not counted in economic terms.

At the end of the war, France, Germany and Great Britain had all the ingredients for a successful aviation industry: by late 1917, air mail services had begun. Their surface transportation systems were destroyed. There were airmen looking for civilian work, surplus aircraft, and venture capitalists willing to initiate airline projects. In addition, the national prestige associated with an airline encouraged governments to subsidise the nascent industry by awarding grants to approved companies (Davies, 1964: 149).

With the inauguration of the Paris–London service in 1919, France became the first European government to subsidise its airlines. Great Britain followed suit, rescuing its airlines from bankruptcy; in 1924, they merged to form Imperial Airways Ltd. Germany subsidised civil aviation during the interwar years more than any other country; by 1926 its airlines were consolidated into Deutsche Luft Hansa.

European states used their airlines as 'chosen instruments' to execute political and economic policies (Nayler and Ower, 1965: 195). In particular, they linked empires. This motivated the development of fast, safe, and reliable equipment, but it led to a myopic perception of aircraft performance requirements: the geographical relationship of the colonies to their mother countries meant the need for increasing range capabilities was ignored.[2]

Because of aviation's strategic value, each state supported only its own airline and jealously guarded such political rights as customs, landing privileges, and route establishment (Davies, 1964: 67–8). The idea of subordinating national pride to facilitate an international air transportation system was inconceivable. This eventually resulted in disorganised competition among too many small airlines.

The early link between aviation and political purpose may explain why Europeans treated the industry as a 'participatory enterprise': state involvement meant more control over industry development than a 'regulatory enterprise' approach would have.

Expanded Horizons

On 28 August 1919, the leaders of six European airlines met at the Hague to form the International Air Traffic Association (IATA) 'to co-ordinate the operations of international air transportation' (Taneja, 1980: 2). Two decades later, IATA had members from North and South America, Asia, and Africa as well. Its early accomplishments were in the field of standardisation.

> Safety in flight . . . and economy in operating costs . . . are two goals towards which the airlines must constantly strive. Safety and economy are unfortunately very often opposed, but standardisation irreproachably conceived and widely carried out – offers a straight way to both of these essential aims at one time. (Brancker, 1977: 7)

Some IATA recommendations seem elementary now, but were revolutionary then. For example, one required all throttle controls be designed so a push forward caused acceleration and a pull back caused deceleration (Brancker, 1977: 7). As a private association, IATA could not make legally binding requirements. Its success was related to the close relationship between airlines and aircraft manufacturers, which were often parts of the same company (Gidwitz, 1980: 38).

Meanwhile, at the 1919 Paris Convention, 32 states ratified the International Convention for the Regulation of Aerial Navigation. The primary accomplishment of this first multilateral treaty concerning air law was the recognition 'that every Power [not just signatories] has complete and exclusive sovereignty over the air space above its territory'.[3] The assignment of air space property rights meant international aviation development 'would have to proceed on the basis of agreements among states, rather than on any general principle of . . . freedom of the air' (Salacuse, 1980: 814).

The second accomplishment was the creation of the International Commission for Air Navigation (ICAN). It met regularly until 1939 and facilitated state co-operation in many technical aspects of civil aviation. Its failures were significant, however: first, states excluded economic regulation of commercial aviation from the provisions of the Paris Convention because they realised its potential and sought 'to protect their own national airlines' (Salacuse, 1980: 816). Second, although it could issue legally binding decrees, it was unable to enforce them because of conflicting national interests and because the United States was not a member (Salacuse, 1980: 831fn115).[4]

In a counter measure, the US formed the Inter-State Aviation Commission in 1923. Five years later it became the Pan-American Convention on Commercial Aviation and had 22 members. Like IATA, it issued only recommendations.

Initial US Government Involvement

On 15 May 1918, the first US air mail route was established. Serviced by government personnel, the state assumed responsibility for mail delivery, and indirectly for pilot and aircraft safety. In these early days, a pilot who refused to fly was not paid, regardless of the circumstances. Most pilots only refused to fly when they were concerned about safety conditions associated with the flight assignment, such as the status of the aircraft or weather. Many felt that airfield managers, especially those who were not pilots, were willing to jeopardise pilot lives in order to ensure timely mail delivery. A three-day strike in 1919 inspired the first safety regulation and dealt with the problem of fog: if a pilot disagreed with a flight manager's assessment of the weather conditions, the latter was required to prove it was safe either by flying the plane or by sitting in the mail bin while the pilot flew (Solberg, 1979: 19).

One of the first safety lessons regarding aircraft manufacturing and human factors was learned in the 1920s: the De Havilland DH4 with a Liberty engine kept burning in flight. Test pilots discovered the long exhaust pipes were so close to the fuel tanks that small amounts of gas spewed on to the hot pipes and ignited. Shorter exhaust pipes eliminated the fire hazard, but at night their flames blinded the pilot: ' "It was my first lesson in human factors," said Jerry Lederer, the pioneer Air Mail engineer who later became a leading safety expert. "I had to take them [the exhaust pipes] off." ' (Solberg, 1979: 23.)

Initially, there was little state-imposed safety regulation of American commercial aviation; airlines were left to design and implement their own standards. In the 1920s and 1930s it was in the interest of the airlines to maintain a safety-oriented reputation to encourage a timid public to fly. But, for economic and competitive reasons, safety first was not always the rule: speed was encouraged above all else, because it was the only real advantage air transport offered over an efficient highway and railway system. Eventually, an increasing number of accidents resulted in the public demand for the creation of some kind of 'authority in commercial aeronautics' (Solberg, 1979: 32).

At this time, the development of aviation in the United States was lagging behind Europe in all respects – manufacturing, service and international co-ordination. The 1925 Air Mail (or Kelly) Act authorised President Calvin Coolidge to appoint an Aircraft Board[5] to determine aviation policy. Its recommendations reflected the perceptions of state elites that both aviation's manufacturing and service sectors were critical to national defence and other policy purposes. Therefore, aircraft manufacturing would be supported by a five-year Army and Navy procurement programme and air transport would be supported indirectly by regulation and air mail subsidies.

The Department of Commerce was charged with full regulatory responsibilities. Its Bureau of Air Commerce 'set up federal licensing standards for planes and pilots', and began to 'establish and maintain flyways' just as the government 'maintained waterways and highways' (Solberg, 1979: 64; Lederer, 1987: personal correspondence). The Postmaster General was empowered to award air mail contracts, which 'covered the added expense of acquiring and operating passenger equipment' (Davies, 1964: 55). Eventually, several small airlines merged to become 'the Big Four': American Airways, Eastern Air Lines, Transcontinental and Western Air (TWA), and United Air Lines. In 1929, when it was discovered mail services were provided more efficiently between Latin America and Europe than between Latin America and New York, subsidies to Pan American Airways were increased and it was given a virtual monopoly over international air transport services.

Two landmark accidents inspired a call for even more rigorous safety regulation. In 1931, a Fokker plane carrying Knute Rockne crashed, killing all aboard. Department of Commerce investigations attributed the cause of the crash to structural failure. The wooden wings had rotted and fallen apart in midair (Solberg, 1979: 155).[6] Four years later, a TWA DC2 crashed and killed the popular Senator Bronson Cutting of New Mexico. This time investigation conclusions led many to believe intradepartmental conflict of interest was preventing a more accurate determination of the accident's cause. Separate Senate and Federal Aviation Commission investigations concluded that a single, independent organisation should be created to manage the airline system, with an autonomous board responsible for accident investigation. The dual responsibility for promoting the industry and regulating its safety had proved incompatible (Solberg, 1979: 149–98; Davies, 1964: 136–7).

The Civil Aeronautics Authority (CAA) was created in 1938 to develop a commercial air transport system to 'meet the needs of the foreign and domestic commerce of the United States, of the postal service, and of national defense' (Solberg, 1979: 200). An independent Air Safety Board, with sole responsibility for investigating accidents was also established. The first accident-free year in US civil aviation history was 1939.[7] In 1940, the Civil Aeronautics Board was created to streamline administrative processes. Its responsibilities included both economic and safety matters.

State elites had come to realise government intervention was imperative to the creation and maintenance of an environment conducive to the development of civil aviation. Left to its own devices the fledgling industry had proven incapable of enforcing the kind of operating discipline needed for further growth. The economic 'laws' of minimising losses while maximising short term gains had led to the incorporation of a speed-(or any other new technology) at-any-cost mentality since that was the primary advantage airlines held over other transportation modes. No single aircraft manufacturer or airline operator could afford to impose costlier safety regulations without a guarantee that competitor compliance would follow; to have done so would have meant certain bankruptcy due to lowered profit margins. There was no mechanism – state, market, or otherwise – to ensure a minimum level of safety; remedies awaited crises. Once the number of accidents reached an unacceptable level,[8] the state responded to 1) domestic pressure, 2) commercial aviation's revenue potential, and 3) sovereignty and national security issues by creating organisations and rules designed to achieve the safety goal.

Still, civil aviation was treated as a 'regulatory enterprise' in the United States, where it was believed that individual calculations of self-interest with minimal government constraints was best for industry growth.

PHASE II: INTERNATIONAL AIR SAFETY ORGANISATIONS FOUNDED, THEN CONFOUNDED BY NATIONAL BEHAVIOUR

Europe's early lead in all aspects of civil aviation was slowly eroded by competition from the United States. By 1928, American airlines flew more mileage annually, and, by 1929, they carried more passengers; by the late 1930s, the United States dominated aircraft manufacturing.

At the end of the Second World War, everyone recognised that air transport was ready to expand far beyond the sporting and military purposes once viewed as its only domain. The general public was now 'air minded', more frequently opting for air transportation when travel distances exceeded 200 miles. This was important not only in the US, with its immense geographical span, but also in Europe where increasing appreciation of national interdependence led to more travel across state borders (Davies, 1964: 238–40, 271–72).

In addition, during the war, nearly a quarter century of technological advancement was telescoped into six years because states were willing to fund the research and development required by the military. Radio communications, navigational aids, and other instrumentation were vastly improved. Speed, range, and reliability were increased. Pilots were better trained and more sophisticated. The pre-war grass airfields were transformed into asphalt or concrete runways. Allied Europe with its war-devastated factories and Axis Europe with its factories immobilised as a penalty for defeat stood to lose as much as the US had to gain from these changes which enhanced safe and efficient commercial operations.

With the development of the airplane into a major instrument of global transportation, the tensions between state goals and global needs resurfaced. The 'freedom of the air' versus 'air space sovereignty' debate was back on the bargaining table, this time more concerned with economic, than political-military issues. States realised they could not solve the problems of international commercial aviation unilaterally. Flight techniques and parochial flight laws needed to be co-ordinated, and technical and safety information easily disseminated. Uniform standards were needed to govern airport design, air traffic control systems, weather reporting, and personnel training and licensing. A forum for settling legal questions related to landing, commercial, and overfly rights was imperative.

ICAO and IATA

In 1943, President Franklin Roosevelt and Prime Minister Winston Churchill met to discuss the possibility of multilateral regulation. The United States agreed to host an international civil aviation conference 'to make arrangements to allow international airlines to develop commercial air transport services' (Taneja, 1980: 8).[9] Many hoped the conference would accomplish for international civil aviation what Bretton Woods had for the international monetary system and postwar reconstruction. It did provide for the establishment of an international forum, the International Civil Aviation Organisation (ICAO), which officially came into existence on 4 April 1947. To achieve its broadest aims, the Conference first had to address intermediate political, technical, and economic goals. Here, it achieved only partial success.

The political goals 'concerned the arrangements for obtaining authority to overfly another nation's sovereign territory and to make stops in foreign territory for technical reasons' (Thornton, 1970: 20).

In the technical arena international arrangements were needed for pilot and mechanic licensing, aircraft airworthiness certification, equipment and regulatory standardisation, data collection and dissemination, and so forth (Thornton, 1970: 20).

> The economic objectives included: the assignment of air routes to nations and to airlines; the arrangement for setting air fares, frequencies, schedules, and capacities; and methods of facilitating interairline fare transfers, customs arrangements, cooperation in servicing and coordination of schedules. (Thornton, 1970: 20)

During the first six weeks, delegates considered four draft conventions. New Zealand and Australia proposed a supranational arrangement in which a world organisation would own and operate all international airlines. From a more moderate perspective, Canada sought an international authority to fix and allocate routes, frequencies, capacity and rates. The British envisaged the multilateral surrender of sovereignty to a similar international regulatory body. However, the US wanted international aviation to be guided by free competition (Jönsson, 1981: 280–81; see also Salacuse, 1980: 819–26).

Eventually the two contending approaches of the major aviation powers – the United States and the United Kingdom – became the focus of debate. To break the deadlock, the Canadian delegation submitted a compromise proposal of four principles to be the 'universally applicable working rules for unilateral air transport relations':

> 1) A civil aircraft has the right to fly over the territory of another country without landing, providing the overflown country is notified in advance and approval is given . . . called 'the right of innocent passage.' 2) A civil aircraft of one country has the right to land in another country for technical reasons . . . without offering any commercial service to or from that point . . . called a 'technical stop.' . . . 3) An airline has the right to carry traffic from its country of registry to another country. 4) An airline has the right to carry traffic from another country to its own country of registry. (Gidwitz, 1980: 49)

Known as 'freedoms', these were not sufficiently liberal for either the UK or the US. Operating from different bases of power – the American domination of aviation's manufacturing and service sectors, and the economic, political and military strength accrued in the aftermath of the First and Second World Wars, and the British control of a number of strategically important geographical locations – both sought the operation of long, global routes, which international carriers required to be profitable.[10] So, a fifth 'freedom' was added to the agenda:

> 5) An airline has the right to carry traffic between two countries outside its own country of registry as long as the flight originates or terminates in its own country of registry . . . called 'beyond rights'. (Gidwitz, 1980: 50)

Following a Dutch suggestion, only the first two freedoms were actually adopted by the conference. Considered 'political' freedoms, these are necessary to operate any service over or through a foreign country. The last three freedoms, essential for the exploitation of the world's markets, are called 'commercial' freedoms.[11]

The conference achieved its political goals primarily because they endorsed the principle of sovereignty which every state wished to keep for military, political and economic reasons. Militarily, states could better protect their borders if they could control when and by whom their air space would be penetrated. Politically, all states gained greater, if imperfect, equity in the decision-making processes associated with air transport by gaining control over the allocation of related values domestically and internationally. States could make decisions regarding who got landing rights into which gateway airport; and they could own and operate their own airline which could be used to implement policies (Krasner, 1985: Chapter 8). Economically, states could generate revenue, employ people, exercise some control over trade and have some say in the development of this strategic industry (Jönsson, 1981; Krasner, 1985; Salacuse, 1980; Thornton, 1970).

The conference achieved only some of its technical goals. It was unable to establish an airworthiness code which could facilitate the trade of aeronautical products.[12] In part, this was due to the cumbersome procedures involved in trying to get over fifty states to agree to a standard.[13] And by the time one was considered acceptable, advances in technology required a new, more sophisticated standard. Third, ICAO had weak enforcement powers. Articles 37 and 38 only required members to comply to 'the highest practicable degree' (Conventional on International Civil Aviation, 1944). Finally, the US refused to accept ICAO standards as its own (Beard, 1982: 10). Foreign manufacturers were eager to enter and compete in the civil aircraft market then dominated by the US, so they turned to the use of Bilateral Airworthiness Agreements.[14]

Conference delegates could not agree on how to structure a regime to govern the commercial aspects of international aviation. This meant their 'post-war management would have to develop on the basis of bilateral agreements between concerned states' (Salacuse, 1980: 826).[15] And it led to the 1945 formation of the International Air Transport Association (IATA) by airline representatives present at the Chicago Conference who wanted an organised airline group to represent industry views at ICAO meetings. Often accused of being a global price-fixing cartel, IATA's stated goals are operational economy and efficiency. Like ICAO, most of IATA's work in the area of safety has been directed toward equipment and procedural standardisation (Brancker, 1977: 18–37).

Bermuda I: Unilateral Management Revisited

The negotiation of the 1946 Bermuda Agreement between the United States and the United Kingdom was important for two reasons. First, it was an acknowledgement that, since states owned the air space above their territories, the only way to manage air transport was with a bilateral agreement which, in effect, enabled states to make unilateral decisions. Any other multilateral management scheme would have undermined state sovereignty too much to be acceptable.

Second, Bermuda I became the model for all subsequent Bilateral Air Transport Agreements. It resolved the debate over treating civil aviation as a regulatory or participatory enterprise which had stalled international air transport. The British feared the former approach would lead to continued US domination of the industry and wished to prevent its becoming a prevailing norm of the international air transport regime. They therefore adopted a position which they knew would be unacceptable to the US; they refused to grant the three 'commercial' freedoms unless London could serve as the gateway to Europe.

Resolution awaited an interesting interplay of economics and politics. The US had superior transport aircraft and possessed the Hawaiian Islands, which could have prevented Britain from linking Australia with Canada. However, the UK had the advantage of potential route structure because of air bases established throughout the empire. The stalemate was broken in 1946, when Britain requested a \$3.75 billion loan in addition to a compromise on the \$650 million lend–lease debt (*US Department of State Bulletin*, 1946: 184). Congress was not in a generous mood, feeling the US had done enough for the British. When the two countries met to negotiate, the UK accepted all five 'freedoms' in exchange for the loan and US acceptance of the IATA rate-setting machinery. These rules guided international aviation for thirty years, but were subject to unilateral demands for change and only as lasting as the shortest duration of one bilateral agreement. The norm of using bilateral agreements to govern international aviation remained.

Both Bilateral Air Transport Agreements (BATAs) and Bilateral Airworthiness Agreements (BAWs) gave the US agency responsible for improving and maintaining aviation safety extraordinary power at the global level. Virtually every state wanted to participate in the largest, richest, and, therefore, most desirable air transport market. BATAs required all airlines to meet US safety standards in order to penetrate US air space or to land and take off from US airports. Similarly BAWs required all aircraft components to meet US quality standards. Thus when US standards changed, so did the world's.

A decade passed, then the United States revised its aviation safety management in response to a public outcry about an increasing number of mid-air collisions. On 30 June 1956, the worst air crash to that time occurred. A TWA Constellation collided with a United DC7 at 21 000 feet above the Grand Canyon killing all 128 persons aboard. Three actions followed: 1) more long-range radar equipment was installed and more rigorous airway traffic surveillance was begun; 2) the 1958 Federal Aviation Act limited the Civil Aeronautics Board to regulating airline economics and competition and created the Federal Aviation Agency (FAA) to promote aviation's economic development and air safety and to investigate air accidents; 3) the new FAA Administrator strengthened enforcement of existing safety regulations (Solberg, 1979: 364–5).

On 16 December 1960, there was another mid-air collision. A United Airlines DC8 with 84 persons aboard and a TWA Constellation with 44 crashed over Staten Island killing everyone. This led to three more improvements: 1) speed was reduced in terminal areas; 2) more sophisticated electronic gear was installed in airports and along airways; 3) pilot and air traffic controller training and procedures were changed to facilitate their working at jet speeds (Solberg, 1979: 365–7).

Throughout the decade, government and public concern intensified regarding potential conflict of interest within the FAA because of its dual responsibility to promote aviation and to ensure a high level of safety. Incorporated into the new Department of Transportation (DOT) in 1967,[16] the FAA Administrator was statutorily delegated authority to regulate aviation safety through the DOT Secretary. The FAA's promotional duties found in Section 305 of its charter were specifically not among the enumerated portions delegated to the FAA; these were transferred to the DOT Secretary. The intent was to separate safety from economic regulation because these can be conflictual goals. The DOT Secretary abrogated this intent immediately by delegating the responsibility to foster air commerce to the FAA (*US House of Representatives*, 1980c: 89–90).

In 1975 the National Transportation Safety Board (NTSB) was created as an independent agency with the responsibility of investigating all transportation mode accidents and determining their cause(s). It also monitors the FAA's safety activity. Its aviation section consisted of the transfer of the Civil Aeronautics Board's Safety Bureau.[17]

The Post-Second World War Aviation Regime

The post-Second World War aviation regime embodied the two constitutive principles of international relations: the political one of state sovereignty and the economic one of property rights (Krasner, 1982: 505). They rested on the assumption that the aviation market would continue to expand dramatically. The passage of time revealed the contradistinctive character ot these principles as they were incorporated into this regime.

The US wanted air transport to be treated as an economic activity in which government activity was limited to a minimalist regulatory role. It wanted market conditions to determine routes, capacities, gateways, fares, aircraft qualities, and so forth. American aircraft manufacturers controlled 85 per cent of the market, and US airlines served the largest and most desirable home market as well as dominating international markets. Policy makers from all corners of the world perceived that such a regime could lead to US dominance of this strategically important industry sector into perpetuity; Americans were pleased, the rest of the world apprehensive.

The national security concerns which originally resulted in the adoption of an air space sovereignty principle prevailed. The most basic regime norms arose from this principle.[18] First, the 'right to conduct airline services would be negotiated in bilateral – not regional or multilateral – agreements' (Lowenfeld, 1975: 40). This, plus IATA's rate setting role, protected non-US airlines from the kind of domination many feared would accompany a global level 'regulatory enterprise' approach. Essentially, this norm meant that, at the global level, a 'participatory enterprise' approach reigned. States could exercise significant control over related political, military and economic issues (Krasner, 1985; Lowenfeld, 1975; Salacuse, 1980).

Second, international air transportation would be offered on a scheduled service basis by airlines 'owned by the state or citizens of the state whose flag they flew' (Lowenfeld, 1975: 38). This allowed for another level of state control, and facilitated the use of aviation as a policy tool (Gidwitz, 1980). For example, civil aviation supplemented military transport and 'made possible a compromise between maintaining massive American troop strength in Europe and relying excessively on nuclear deterrence' (Lowenfeld, 1975: 41). This also meant that, domestically, states were free to implement the state–industry approach with which they felt most comfortable ideologically. The United States could adopt a 'regulatory enterprise' approach and keep government intervention to a minimum, while other states could pursue a 'participatory enterprise' approach and more actively promote domestic aviation industries. At the time, it was not apparent these might eventually conflict at the international level.

Third, airlines would compete as a group against other modes of transportation, more so than among themselves (Lowenfeld, 1975; see also Eddy *et al*, 1976).

PHASE III: REGIME CHANGES SOUGHT

While there is still a global consensus that aviation and its safety need to be managed, there is growing dissatisfaction with the current regimes. Concern about the international civil aviation regime reached a peak in 1976 when the British denounced the 30-year-old Bermuda Agreement and demanded its renegotiation. The UK sought more control over route and landing rights, capacity, and prices. The US decided to use the negotiations to effect regime changes. Apparently American policy makers did not see the fundamental juxtaposition of political sovereignty and market operations. Hoping to generate more traffic for US airlines, they gave up some fifth freedom rights, accepted limitations on designated airlines, and allowed controls on price and capacity. They intended these concessions to be a trade-off for more competitive practices by other states' airlines. Other states took advantage of freer access to the US market, but gave up little in return.

Other factors motivated US compromise on these issues as well. European co-operation in other trade arenas and in NATO arrangements were sought (see Haas, 1980: 371–2; Hartley, 1983). European governments were 'becoming less satisfied, not because the imbalance [was] getting worse, but because it [was] not getting better fast enough' (quoted in Feazel, 1984: 135). Compromise was also a function of more equitable (if still asymmetrical) distribution of power resources among a burgeoning number of states. Lack of access to enough key bargaining chips makes it difficult to impose regimes which may not be considered desirable by other states.

The present situation has been affected by surplus capacity and regime linkages. It has been aggravated by the incompatibility between domestic and international approaches to state–industry relationships in commercial aviation.

> The current debate rests on a profound and fundamental conflict between the concept of air transportation as a business which relies on market forces and competition on the one hand and that of air transportation as an internationally regulated industry in which the nations of the world share equitably on the other. (Salacuse, 1980: 837)

Surplus capacity in air transport emerged in the 1960s when the market started to grow at a slower rate than anticipated. In 1962, European states requested a substantial fare rise within the International Air Transport Association to maintain revenues despite the declining traffic. The US opposed the request but was outnumbered in the final IATA tally. The US was able to keep fares low by allowing charter carriers to proliferate and liberalising their rules; this also increased capacity. Technological advances in the 1970s led to the introduction of wide-bodied jets, further expanding capacity. American disengagement from Vietnam freed even more capacity as charter carriers which had flown troops entered commercial service. By 1975, 'unused capacity on the North Atlantic alone was equivalent to 15 000 empty Boeing 747 round trips' (Jönsson, 1981: 286).

Aviation's linkage to other issue areas aggravated the problems associated with surplus capacity by simultaneously increasing costs and decreasing demand. For example, the energy crisis in the early 1970s raised airline operating costs by 10 to 30 per cent and brought on a recession which significantly reduced passenger demand.

Corporate responses were predictable according to Mancur Olson's counterintuitive logic of collective action (1965). Airlines constituted a large group, therefore no major airline curtailed its services to more appropriately match demand 'lest its competitors capture a greater share of the traffic' (Lowenfeld, 1975: 44). Eventually the problems of surplus capacity in the service sector became the problems of the manufacturer as airlines began cancelling existing orders and options and ceased placing new orders. The smaller group of aircraft manufacturers – Boeing, McDonnell Douglas, and Lockheed – adjusted by shrinking production schedules and furloughing employees.

State adjustment strategies have been almost equally predictable. The United States was particularly sensitive to the negative effect these events had on the ability of commercial aviation to generate income, because its economic actors dominated the industry. Operating within an international regime based on 'authoritative resource allocation' inhibited the ability of US airlines and aircraft manufacturers 'to realise fully the economic advantages generally associated with technology, skilled personnel, and large size' (Krasner, 1985: 5, 203, Chapter 8). American policy makers are trying to impose at the global level a market-oriented aviation regime more compatible with the 'regulatory enterprise' approach used domestically. This time they are motivated by a sense of alarm about the US treasury's negative balance sheet. Their perception that civil aviation generates significant revenues vital to an improved domestic economy parallels their belief that it could earn much more if service restraints were lifted internationally. This also dovetails with the liberal notion that economics and politics ought to be separate.

The first step in this process was the economic deregulation of US airlines to make them more efficient, productive, and better able to take advantage of the comparative advantages associated with technology and size. Carriers responded by rationalising production through merger, acquisition and bankruptcy. The short run effects proved more damaging than policy makers had anticipated, nevertheless the long run payoff of stronger airlines to compete internationally may soon be in place (Jönsson, 1981; Golich, 1984; National Academy of Sciences, 1986). Economic deregulation has had a negative effect on safety (see Chapter 4; Golich, 1988; O'Brien, 1987).

Now the US is trying to deregulate international air transport. This effort conflicts directly with the principle of state sovereignty and has met with strong opposition from both advanced industrial and developing states more comfortable with a 'participatory enterprise' approach.

The Europeans argue for continued state involvement because 'international carriers operate in an environment where the principle of sovereignty reigns' (Jönsson, 1981: 288; 1987: 39). National flag carriers do not have the option of merging or going out of business like domestic carriers, so subsidy wars may ensue. The rest of the world fears economic deregulation will 'lead to a redistribution of aircraft capacity away from the less and to the more profitable routes, thinning out services to the Third World in particular, and aggravating the already serious congestion at popular destinations' (Jönsson, 1981: 288; 1987: 36-9; see also Krasner, 1985: Chapter 8). There is a universal concern that ' "survival of the fittest" in international aviation inevitably entails "survival of the fattest", and . . . that the American preaching of laissez-faire really means "laissez-nous-faire" ' (Jönsson, 1981: 289). There is also concern that the accompanying denigration of safety may lead to catastrophes like those of the DC10 crashes in 1974 and 1979 as detailed in Chapter 4.

The Third World demand for a new international economic order based upon 'equitable sharing of economic activities' constitutes an issue linkage which muddies the management of international civil aviation (Salacuse, 1980: 836; see also Krasner, 1985: Chapter 8). With more and stronger states participating in the international commercial air transport regime, it is conceivable that the debate may change from one regarding equal opportunity to compete to one regarding the right to 'an equal share in the benefits of civil aviation' (Salacuse, 1980: 836). The most fundamental concern among current regime participants is that international economic deregulation would mean a change in the basic principle which has guided the regime from the beginning: political sovereignty of air space.

Nevertheless, evidence suggests that Europe may be succumbing to US pressures for global deregulation. A 1984 review by the UK's Civil Aviation Authority concluded 'there is a strong case for removing controls on domestic fares and increasing competition on the routes within Britain . . . the British government is continuing to press the European Commission for a liberalisation of air services within Europe' (European Report: 9). The UK is privatising its air transport sector. It sold British Airways to private investors and is allowing new airlines to enter domestic and international service markets. In Europe, only Portugal and Italy continue to restrain air transport services by intergovernmental agreements. Globally, the US, Canada, the United Arab Emirates, Qatar, and Bahrain have lifted restraints on services ('British Government . . . ,' 1987: 36). Even in South America where Krasner argues states have benefited from their control over air transport services (1985: Chapter 8), several countries are liberalising their commercial air transportation policies in an attempt to attract more outside business (Kolaim, 1987: 133).

Europe is adjusting to US dominance of commercial aircraft manufacturing by co-operative production of its own highly competitive commercial aircraft (see Chapter 3). Between 1980 and 1986, US manufacturers accounted for 80 per cent of the total value of orders for commercial transport aircraft; Airbus Industrie accounted for 15 per cent; Fokker and British Aerospace accounted for the remaining 5 per cent. In the wide-body market, US companies captured 75 per cent, while the Airbus share was 25 per cent. This approaches their originally established goal of 30 per cent (United States Department of Commerce, 1988; Feazel, 1984a, 137).

A second European adjustment strategy has been the development of its own certification standards. This process began in 1960, when 19 European countries concluded a Multilateral Agreement Relating to Certificates of Airworthiness for Import Aircraft. It is now embodied in the Joint Airworthiness Requirements (JARs), a consortium formed by European states which generates its own safety and quality standards. The Europeans have pursued this strategy as a means of avoiding possible market manipulations by the US use of standards (McKeown, 1983: 77; Golich, 1984: Chapters 5 and 6). Chapter 4's discussion of the DC10 crashes will clarify just how vulnerable European states were to US policy decisions. Still this consortium symbolises the remaining tensions between unilateral and multilateral management: on the one hand, its very existence demonstrates European Community recognition that regional states and corporations will have to co-operate and collaborate if they are to compete effectively against US manufacturers and airlines. On the other hand, the participating states have not been able to achieve a legally binding status for JARs because of lingering state resistance to yielding their final authority over the industry (Kennedy, 1979; personal correspondence; Koplin, 1987: personal interview).

The interplay of domestic pressures and trade-related issue linkages is acute. Europeans, worried about lagging economies, pressure their governments to stimulate economies with protectionist moves such as retaliation barriers, subsidies, and 'buy Europe' policies. Despite objections from some sectors, the general consensus is that protectionism 'does work in the short-term. It pays very quick and very large political dividends. And it doesn't cost a thing' (quoted in Feazel, 1984: 137). For state elites, protectionism 'can be compelling simply because it lies within the power of an individual nation' (Cowhey and Long, 1983: 162). Still, Europeans are keenly aware of the importance of the US market for the sale of commercial class aircraft and recognise the US ability to close its doors in a manner so subtle as not to violate current GATT stipulations that trade in aeronautical goods be as free as possible (Golich, 1984: Chapters 5 and 6). Many Europeans fear the US will retaliate and use standards or other non-tariff barriers in a trade discriminatory manner.

CONCLUSION

In analysing the evolution of the international commercial aviation safety regime, it is tempting to argue that air transport should be treated as a collective good. Much of the literature implies that international regimes were created to manage collective goods type issue areas[19] on a scale approaching global proportions.

Adam Smith and Charles Kindleberger make the strongest arguments for treating international air transport this way. In 1776, Smith averred that one of government's three functions is 'to erect and maintain those public institutions . . . such as roads, bridges, canals, and harbours' and to construct 'public works such as docks and domestic transportation systems' (1776: 653, 669, 681-82) which would facilitate free trade (Gilpin, 1977: 21; see also Burnham, 1982; McDonnell, 1982; Miller, 1982; and Siegen, 1982). Initially, this meant roads were passable and oceans were free of piracy. In civil aviation, this means that airports, skyways and aircraft are safe and mutually acceptable rules facilitate commercial transport between destinations.

Whether or not air transport should be treated as a public good became a moot point as soon as property rights were extended over the territories of sovereign states. Susan Strange and John Ikenberry suggest theoretical explanations for why air space property rights were established. Strange argues that 'security from external attack and the maintenance of internal order are and always have been the first concern of government' (1985: 36). Any politically organised society must constantly balance four values – wealth, order, justice and freedom – to survive. The provision of security is paramount to the performance of those tasks.

Carrying this contention a step further, Ikenberry asserts that survival requires state elites to 'harness domestic wealth', because this strengthens the state's foreign position (see also Gilpin, 1975). The challenge is to do so while fostering economic growth and commanding loyalty from the capitalists who generate the wealth (1986: 55–6). The ensuing compromise was that states would provide security and a stable economic environment in exchange for society's revenue creation and its establishment of those parts of the transportation infrastructure which were amenable to generating private profits and sufficient economic efficiencies simultaneously.

The state's need to provide security from external attack was clearly influential in the initial extension of property rights into the skies. The prevailing technologies and knowledge were also significant factors in this process.[20] Skies were 'protectable' at that time because the vehicles then capable of penetrating air space flew within easy range of available weapons systems.[21]

The establishment of property rights created economic interests to be protected. All technologically capable states have consistently pursued strategies of support for civil aviation. From time to time, the desire to unilaterally promote a healthy domestic civil aviation industry has conflicted with the global requirement for co-operation in the co-ordination of route and capacity as well as safety considerations.

Thus the establishment of air space sovereignty set the stage for states to be individually interested in how the commercial aviation regime would be structured. This manifested itself most clearly when participants in the 1944 Chicago Convention insisted the principle remain intact. As a result, states retained their ability to use civil aviation to serve political and economic policy purposes. It exemplifies Chris Mason's contention that

> A condition of interdependence tends to compel states to discuss, and make agreements about, how each is to regulate a wide range of economic activities within its own territory as well as how each is to regulate international trade in order to ensure that the results

> of the transactions which flow between them are as consistent as possible with what each is trying to achieve within its own territory. (1979: 3)

Only the US and the UK had enough power resources at the end of the Second World War to be able to effectively bargain for the remaining elements of the regime. Poised to benefit tremendously if the rules were 'right', and holding valuable bargaining chips, the US created a regime to best serve its interests, with a few modifications extracted by the UK. These great powers instituted a regime which was 'to their liking that [would] not prove totally unacceptable to key partners' (Cowhey and Long, 1983: 160fn5). Once in place, the regime dictated the behaviour and scope of options available to those who wished to participate. The regime worked well until recently. Now the various actors are seeking changes so their interests might be better served.

Aviation safety has not been perceived as an issue whose resolution was of paramount concern. It has been considered important because unsafe airways, airlines or airports diminish the potential of the industry to serve political and economic ends. To serve these functions, the general public and business must be convinced of air transport's reliability, and that requires safe operations. Thus, safety generally has been subsumed within the broader context of the industry's vital role in the further development of communication and transportation systems.

During the transition periods of regime changes, various adjustment strategies may be tried by state and non-state actors. It remains to be seen whether international aviation follows a US model of economic deregulation. The events outlined in this chapter reveal the emerging tensions between the state sovereignty and property rights principles as they were integrated into the international commercial aviation regime. States were ultimately given the right to participate as much or as little as they cared in aviation decisions. The US adopted a regulatory-enterprise, market-oriented approach domestically. This worked well at home and globally because there were few effective international competitors. Most other states intervened more actively in industry decisions (some with more positive effects than others). The disparities between these two approaches were not immediately apparent because the primary markets were domestic. As international markets became increasingly important for both goods and services, the sovereignty versus property rights tensions intensified. If the market is global and structured to minimize state involvement – which a move to international economic deregulation would foster – then the state becomes vulnerable to corporate strategies which may conflict with state goals. Chapter 3 begins to explain how significant a norm change regarding air space sovereignty would be.

3 The Value and Dynamics of Civil Aviation

> What we have is the strangest of paradoxes: a global enterprise boasting some of the world's most advanced science and engineering that operates according to a set of anachronistic, mercantilistic rules consciously crafted to impede efficiency and limit opportunity. It is the marriage of 20th-century technology to 18th-century economics. (Jeffrey N. Shane, Deputy Assistant Secretary for Transportation Affairs, before the Wings Club, New York City, 26 February 1988.)

Civil aviation is strategically valuable to states fortunate enough to participate substantially in the industry. This is a centripetal force impelling states and corporations to dominate the industry unilaterally as much as possible. Each actor wishes to control the allocation of the political, military and economic benefits it offers. The industry's dynamics make it a 'high-risk game with few winners and many losers' (Bacher, 1984: 8) and constitute a centrifugal force inducing political and economic actors to internationalise all levels of production and distribution in an effort to minimise losses and maximise gains by spreading risk and increasing market access.

Analysis of civil aviation's value and dynamics demonstrates the unique position states occupy 'at the intersection of domestic and international systems.' It affirms John Ikenberry's conceptualisation of states as organisations 'staffed by officials whose positions are ultimately insecure' (1986: 54, 61) and who are responsible for making and implementing policy. A state's preferred policy is a function of both its place in the international system and the perception by state elites of what constitutes appropriate behaviour. The former is determined by the institutions of and distribution of power within the international system, the latter by the prevailing ideology which delimits decision-maker notions regarding acceptable state-society and state-economy relationships. Policy choices vary according to 'national differences in the organisation of industrial production and the political institutions that surround the state and the economy' (Ikenberry, 1986: 64fn22; see also Katzenstein, 1985). 'The preference function predicts what states will seek to achieve, structural constraints will determine what is possible' (Ikenberry, 1986: 65).

In addition, analysis of civil aviation's value and dynamics highlights the significant role played by corporations in creating or altering both domestic and international structural constraints as airlines and aircraft manufacturers seek to serve their relevant markets. The commercial class aircraft market is increasingly global. Corporate survival strategies of internationalised production may impel states to act in opposition to what might otherwise have been a preferred policy (Golich, 1987).[1] As economic deregulation of American and European airlines proceeds, a similar situation may arise in civil aviation's service sector.

Chapter 2 revealed how actors with relevant power capabilities create international regimes to advance their own interests which 'will not prove totally unacceptable to key partners' (Cowhey and Long, 1983: 160fn5). The following demonstrates how the daily operation of actors within these regimes redistributes power and wealth in a constant 'process of disequilibrium and adjustment.' Eventually a 'disjuncture between the existing social system and the redistribution of power' occurs and an imperative to replace the governing institutions emerges (Gilpin, 1981: Chapter 1; 1975: 21-2). Then, the actors who are most dissatisfied and who have the requisite power capabilities try to transform the international environment into one more beneficial to them (Axelrod and Keohane, 1985: 253; Gilpin, 1981: 9).

The post-Second World War international commercial aviation regime reflected what US policy makers perceived to be in America's best interests. This included incorporating a means of distributing costs and benefits to encourage participant compliance. As a result, relative positions of power shifted. The US remains 'the most powerful actor in the world political economy' and has 'greater leeway for autonomous action than other countries' (Keohane, 1984: 26), but no longer dominates as it once did. Other states are in a position to change regime rules and decision making procedures, if not its norms and principles, so they will receive more benefits than presently.

The first section of this chapter examines the nature of civil aviation's strategic value to states, including its potential as an employer, its generation of revenue, its source of knowledge and high technology research, and its synergistic relationship to national security. This value is reflected in the shelters and supports provided by governments in their efforts to promote the industry.

Civil aviation's strategic value can affect safety. Substantial participation in the industry ensures accrual of a proportional amount of its many benefits. Certification, a state function, is the process used to guarantee the safety of aviation equipment. Historically, the state (or group of states) which dominated the industry also controlled its safety regime. This virtually assures certification of its aviation products and, thus, their access to commercial markets. A potential conflict of interest arises when state elites, responsible for certification, are more concerned with making their state's aerospace products available to the market than with ensuring their quality. The probability for such a situation is increased if elites perceive their state's power position to be eroding and seek to regain a dominant position. This can be myopic: if the products fail to perform according to the guarantees inherent in their certification, then other states may be motivated to increase their participation in the industry by producing of their own equipment and establishing their own certification processes. The relationship between the industry's political-economic significance and the issue of safety is more apparent to state and corporate elites as competition for market dominance intensifies.

The chapter's second major section examines the dynamics of civil aviation. Airline service and aircraft manufacturing are volatile industries, sensitive to economic and political vicissitudes. They are mutually dependent for their success or failure. Manufacturers need airlines to purchase aircraft; airlines need manufacturers to produce aircraft that appeal to the ultimate consumer – the travelling public.

The synergy between industry value and dynamics helps explain what might otherwise be enigmatic policy decisions. The simultaneous operation of centripetal and centrifugal forces helps explain why policies are not always consistent, though the goal of substantial participation, if not industry dominance, is clearly articulated and logically perceived by policy makers.

STRATEGIC VALUE

Economic Issues

The aerospace[2] labour force in the US is significant in terms of both numbers and quality. Aerospace employment has increased at an average annual rate of 5 per cent since 1983. In 1986, the industry employed 1 207 000 persons; of these 401 000 were skilled blue collar production workers, 162 000 scientists and engineers, and 60 000 technicians (Pascale, 1987: 5). It is 'the largest among employers of skilled blue-collar machinists, engineers, and technicians' (Bluestone *et al.*, 1981: 3).

Airlines employed another 421 686 persons in 1986 (*Air Transport 1987*, 1987: 12).[3] Indirect employment in related service industries such as food services, interior design, tourism, fuel management, advertising, and airport facilities construction adds significantly to these numbers.

The majority of aerospace employees are highly skilled and well educated. Referred to as 'technological density' (O'Conner, 1978: 17), this work force can be tapped for various national production needs. In the development of its 757 and 767 air transports, Boeing 'rented' talent from other domestic companies (Bluestone *et al.*, 1981: 129). Without this available employee pool, Boeing might have exported jobs overseas or been forced to cancel the transport programmes.

Since the late 1950s, aerospace has been the leading industrial contributor to the US trade balance. In 1986, aerospace exports and imports achieved record levels of $19.7 billion and $7.9 billion, respectively, for a positive trade balance of $11.8 billion. The positive balance was reached on the strength of civil exports, as military exports declined by nearly $1 billion (Yager, 1987: 1-2).

The European Economic Community (EC)[4] aerospace industry also generates significant revenues. In 1985, the industry earned a positive trade balance of approximately $1.1 billion. About 30 per cent of EC aerospace turnover[5] was exported; 40 per cent were civil sector goods, 27 per cent military ('Continued International. . . ' 1987: 5).

Airlines also contribute positively to a state's balance of payments. According to Robert Thornton, even with heavy government subsidies, 'Airline operations produce important contributions to the foreign exchange earnings of the operating nation' (1970: 100).

The aerospace industry plays a critical role in creating and maintaining a strong knowledge and technology base. In 1978, the National Academy of Sciences Committee on Technology and International Economic and Trade Issues concluded that a state's competitive position in world trade is a reflection of the conditions of its domestic economy. It further concluded that industrial innovation was a major factor affecting a healthy domestic economy. More recently, the same committee determined that civil aerospace technologies are vital to the total constellation of US technological leadership. Technology-intensive industries[6] are reserves for information and discoveries which can positively affect other industry sectors. Market (and military) requirements drive the industry to stay at the leading edge of technology. Aerospace is a 'pioneer' industry where new technology is applied early. Experience and usage eventually decrease costs, facilitating diffusion to other industries, manufacturing processes, materials, and research methods. For example, the search for lighter but stronger materials for airframes was driven by the need to cut fuel costs by saving weight. The composites discovered are currently used in automobiles, boats, rapid transit vehicles, and a variety of sports equipment.

Advancing technology increases labour productivity and brings a higher quality product to the market. From 1960 to 1976, aircraft labour productivity increased by 65 per cent (Bluestone *et al.*, 1981: 145). In 1969, 25 000 workers were needed to assemble seven Boeing 747s per month; ten years later, 11 000 workers were able to produce the same number of aircraft ('Masters of the Air', 1980: 54). Technological sophistication generally boosts product salability.

> Conceptually, a superior technology embodied in a particular product is equivalent to a tangible surplus value which the buyer gets . . . in the form of the product's greater economic efficiency compared with similar products embodying less advanced technology. (Boretsky, 1975: 74)

Skills developed in spin-off industries can be important to a state's economy as well. Successful marketing strategies may allow manufacturers to penetrate formerly inaccessible markets. Britain's Dowty Aviation Division, a private manufacturer of aviation components and subsystems for both civil and military purposes, broke into the US aerospace market using tactics learned while collaborating with Boeing and McDonnell Douglas ('UK Firm Uses . . . ', 1977: 159).

Research and development (R&D) outlays offer a tangible measure of an industry's focus on technological advancement. Here, aerospace is a perennial leader among US industries. In 1985, it spent $18 billion, which amounted to about 23 per cent of all US spending for industrial R&D. When measured as a percentage of sales, aerospace also leads all US manufacturing industries. In 1984, aerospace company funding for R&D (exclusive of government funds) equalled 4 per cent of net sales, compared with an average for all manufacturing industries of less than 3 per cent. Combined with government funds, aerospace outlays reached 17 per cent of sales, compared with the all-industry figure of 4 per cent (*AIA Aerospace Facts & Figures, 1986/87*, 1986: 112).

National Security Issues

Obviously aerospace is important to a state's security interests because it comprises the building of missiles, military aircraft and other defence-related materials. Less obviously civil aerospace is also vital to national security. Chapter 2 explained how, initially, military and civilian sector research and development goals overlapped. While this is no longer the case, a synergy remains. Commercial design and production teams can shift their attention to military hardware development. Market requirements often stimulate technological and product advances relevant to military needs and vice versa. Thomas J. Bacher, Director of International Business for Boeing Commercial Airplane Company, suggests that military aerospace products normally precede commercial aircraft in the industrial development sequence, but once the industry has reached a mature stage, a mutually beneficial technology interchange occurs (1984: 6). Figure 3.1 shows this relationship graphically.[7]

A less tangible measure of the strategic value of commercial aircraft production is found in its role as a major purchaser of goods and services from a broad range of commodity, manufacturing and service industries (see Figures 3.2 and 3.3). It supports a production base of 15 000 firms. Significantly, aerospace industries are heavily dependent on electronics, communications equipment and scientific instrument industries, which explains why technologically advanced countries with the requisite industrial infrastructure dominate the sector (Loren and Yager, 1987; see also Neuman, 1983). Civil aircraft manufacture accounts for 80 per cent of the total aircraft production weight during times of peace. Together, the availability of personnel and production base reduces the cost of providing an essential military industrial base and wartime mobilisation surge capacity (see *The Competitive . . .* , 1985, 1–2, 25; *National Aeronautical . . .* , 1986; Bacher, 1984: 6).

Figure 3.1 Military–civil Technology relationship

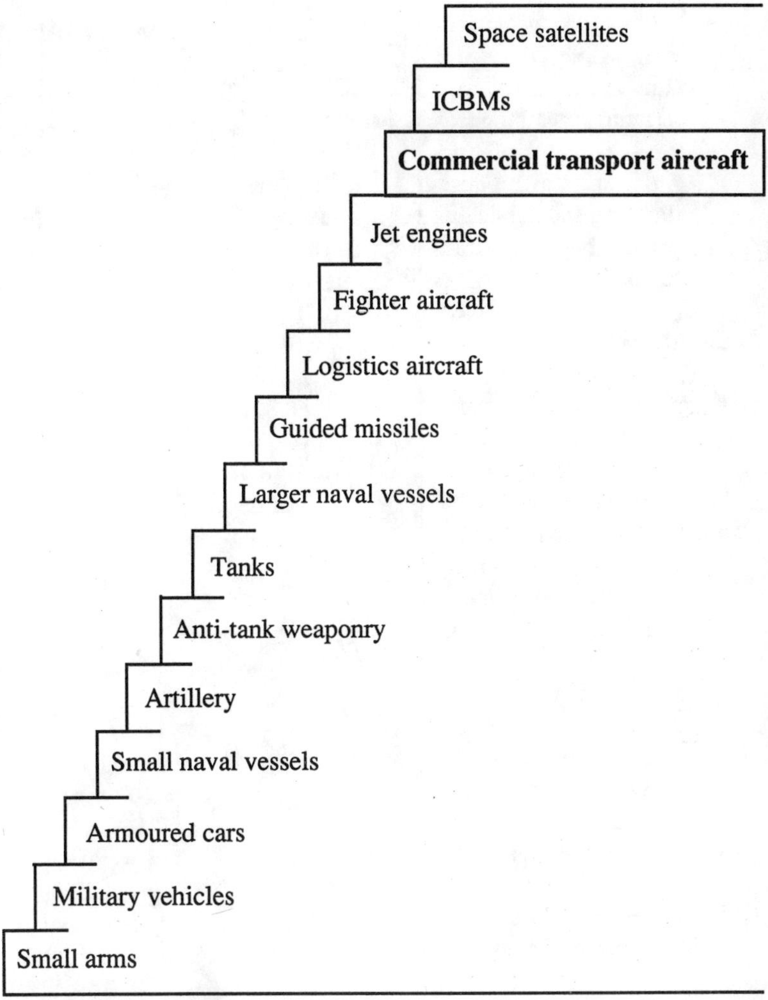

Source: Bacher, 1984:6

Civil aviation's service sector comprises 'a fleet of planes, supporting equipment, and a staff of trained personnel which can be used in flying men and matériel to a combat theater' as was done by American carriers during the Second World War, Korea, and the Vietnam conflict (Stratzheim, 1969: 15). Commercial airline personnel serve in military reserve units.

Figure 3.2 Aerospace industry suppliers

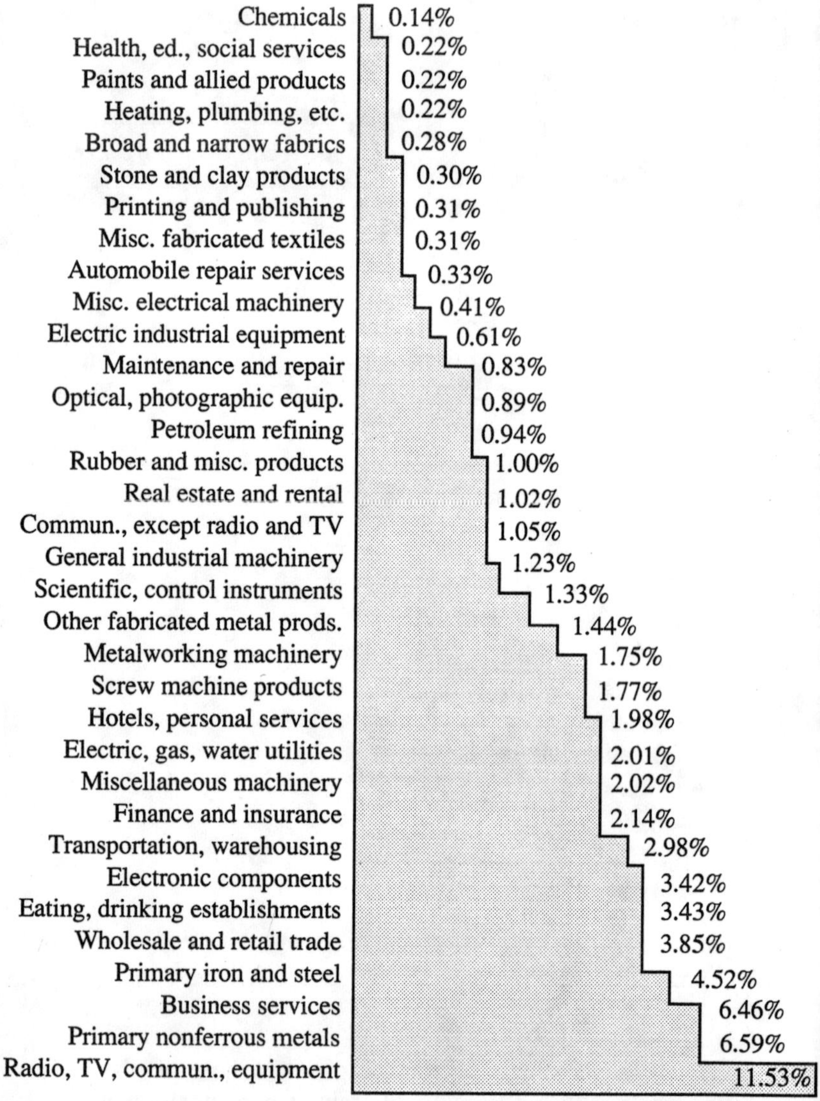

Not shown: Intra-industry purchases, 31%

Source: US Department of Commerce, Bureau of Economic Analysis, *The Detailed Input/Output Structure of the US Economy.*

Figure 3.3 Percentage of total output purchased by the aerospace industry

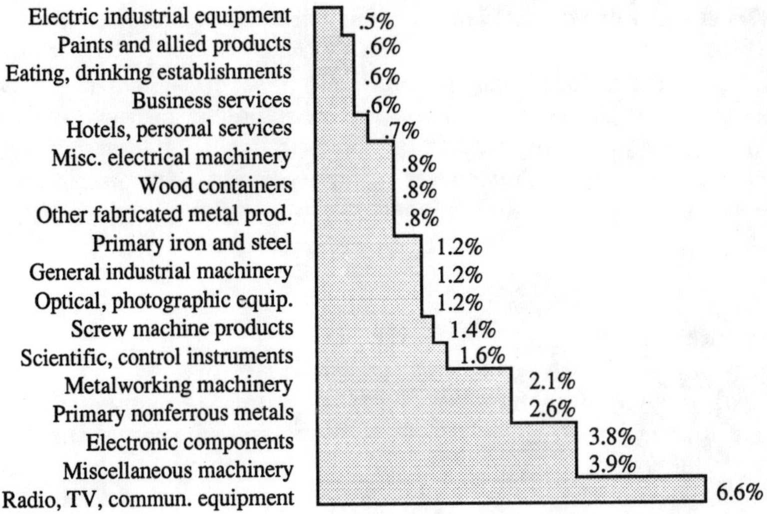

Shows the relative importance of aerospace purchases to the named industries

Source: US Department of Commerce, Bureau of Economic Analysis, *The Detailed Input/Output Structure of the US Economy*.

The military requires 'a strong, profitable, technically up-to-date commercial airline industry' (Thornton, 1970: 80). This is compatible with industry objectives and means the air transport infrastructure must be safe, sophisticated and efficient.

When military needs are less compatible with industry goals, states subsidise commercial carriers. The Civil Reserve Air Fleet (CRAF) in the United States is typical. It confers on the state contractual authority to use certain aircraft from a commercial airline fleet for military purposes. In exchange, the state 'provides added spare parts backup and supports communication facilities it would not otherwise have established' (Thornton, 1970: 80-1; see also Thayer, 1965: 123).

Airlines can also be used as instruments of combat and espionage. During the 1968 Soviet invasion of Czechoslovakia, Aeroflot-marked aircraft filled with military personnel took control of major air fields. International airlines offer 'cover employment' for home state spies. Nominal commercial carriers can conduct espionage: the former Air America and Continental Air Services, used in

South East Asia during the Vietnam conflict, were funded by the US Central Intelligence Agency, the Agency for International Development and other independent executive agencies (Gidwitz, 1980: 26-8).

Finally, a well-honed commercial air transport system provides a fundamentally sound national transportation infrastructure and a vital communications network domestically and internationally. Domestically, it facilitates direct interface among government and business personnel. Internationally, it establishes vital links with allies.

Political Factors

Civil aviation has often been used as a tool of both domestic and foreign policy. Europeans used airlines as political instruments from the very beginning. They consciously developed airlines and routes to provide communication and transportation links between colonies and the mother country. A domestic civil aviation industry serves several political purposes.

State elites consider both the service and manufacturing segments of civil aviation to be prestigious.[8] International airlines prominently display the home country's flag and are known as flag carriers. They symbolise a state's technological advancement and related economic and political prowess. Similarly, participation in commercial class aircraft production imparts the notion that a state has achieved an impressive level of technological sophistication. Thus, some relatively resource-poor states will try to sustain this industry in spite of its dependence on a large and complex high technology infrastructure. The anticipated payoff is continued participation in the industry and the reaping of associated benefits.

Domestically, civil aviation can strengthen a state's economy, and thereby its overall power resource base.

International airlines are used to execute foreign policy because they are inherently mobile, flexible and span 'a wide geographic sweep' (Gidwitz, 1980: 22). For example, bilateral air transport relations between the Soviet Union and the United States were originally influenced by the politics surrounding the East–West split of Germany and Berlin; they continue to be on-again, off-again affairs depending on USSR behaviour in Poland, Afghanistan, and in world affairs generally. Another classic example occurred in 1977; the United States granted an operating permit and landing rights for a joint Syrian–Jordanian air service to New York over the objections of aviation advisers because the US was concerned with the progress of the ongoing Middle East peace negotiations (Gidwitz, 1980: 24).

States may require their airlines to purchase foreign-built aircraft to gain favour with that country, even if they do not fit airline needs. In 1978, Japan's foreign policy bureaucracy coerced Toa Domestic Airlines (TDA) to buy the European Airbus to calm European Community concerns over its large trade deficit with Japan. The airline preferred the McDonnell Douglas DC10 and the US was also experiencing a large trade deficit with Japan. 'Several European governments applied greater pressure and the Japanese government responded accordingly' (Gidwitz, 1980: 26).

Supports and Shelters

The special relationships between states and civil aviation reflect the industry's perceived value. States have offered rhetorical, financial and organisational support to encourage and maintain a healthy domestic aviation industry. Policies enunciated and implemented by US presidents, from Woodrow Wilson to Ronald Reagan, have been designed to support civil aviation.[9] Most recently, manufacturing sector policies have included increased federal support for aeronautical research and development and the relaxation of antitrust regulations to make joint production arrangements more feasible. In the service sector, economic deregulation of American airlines was initiated in 1978 in a conscious attempt to strengthen US airlines as international competitors.

European countries, individually and collectively, also recognise aviation's strategic value. One of the most important statements reflecting this perception emerged from a study conducted for the Commission of the European Communities in 1975.

> The Governments of the Community are at a cross-roads. If they . . .continue to pursue divergent national policies, this could lead to the disappearance of an autonomous European aircraft industry thus damaging the economic, political and social future of the Community. ('Action Programme for the European Aerospace Sector' 1975: 13, as quoted in Edmonds, 1978: 9)

In the service sector they are cautiously pursuing similar policies to maintain or increase their presence in these markets (see Jönsson, 1981; Golich, 1984; Greenhouse, 1987: D1,D7; 'Common Market Threats . . . ', 1987: 146–48; Heard, 1987: 54).[10]

Shelters are 'any market entry barrier provided by some level of government that insulates an industry or firms within it from normal competitive pressures.' The three major categories of shelters for the aircraft industry are 'research and development funding, government-subsidised plant and equipment, and contracting procedures' (Bluestone *et al.*, 1981: 157).

Traditionally government funding for aerospace research and development has been high. The government thus assumes the role of product improvement guarantor. In the US and Europe R&D spending usually approaches one fifth of total industry sales (*AIA Aerospace Facts & Figures, 1986/87*, 1986: 112).

Another shelter is the provision of physical capital. In the United States where the philosophy of free enterprise prevails outright public ownership is unthinkable. Still, as a result of military needs during the Second World War, the federal government constructed and then leased plant production facilities to ensure rapid expansion. This, coupled with temporary depreciation allowances, encouraged the aerospace industry to meet production quotas. At the end of the war, facility leases were extended. In 1972, the government still owned 189 plants (Bluestone, *et al.*, 1981: 162).[11] In Europe, publicly owned industry is commonplace, and the aerospace and air transport industries are nearly always government owned or heavily subsidised.

Special contracting procedures promise lucrative profits. Although such shelters in the US are generally awarded only for military programmes, they can still positively affect civil programmes. Contracts awarded for products with civil applications support companies with direct profits and lowered production costs. Contracts for strictly military hardware free internally generated, fungible monies for application in the company's commercial sector. State elites are well aware of this aspect of procurement policy. In the Capitol Hill debate regarding the Pentagon's purchase of airlifters for its Rapid Deployment Force, most of the arguments revolved around questions of economics and which commercial aircraft manufacturer needed the work most desperately; the comparative characteristics of the aircraft were hardly mentioned ('A bitter clash . . . ', 1982: 91–2).[12]

European countries generally award similar contracts. They may also guarantee the sale of domestically produced commercial aircraft by requiring their purchase by state-owned airlines. Both British Airways and Air France receive substantial state subsidies to operate the commercially nonviable Concorde supersonic transport. In 1974, Air France management wanted to replace obsolete French Caravelles with Boeing 737s. The French minister of finance threatened to dismiss the airline board of directors, if they did not wait to 'purchase French-made Mercurers, an aircraft which never went into series production'. Finally, in 1978, the French government allowed Air France to purchase US aircraft and agreed to subsidise the airline 'for losses accruing from Caravelle operations until that fleet could be replaced' (Gidwitz, 1980: 21).

The US has applied indirect pressure to assure American-made product purchases. In 1968, when Eastern Airlines and Trans World Airlines bought Lockheed L1011s with Rolls Royce engines, there was evidence their final commitment depended on Rolls Royce being able to guarantee a countervailing order for a similar complete aircraft package from a British company ('Lockheed Takes the Lead', 1958: 28). More recently, pressure on US airlines to 'Buy American' was manifested when Eastern bought the European Airbus A300. Congress, trade unions and others lobbied heavily against the purchase ('Dollar Slide . . . ', 1978: 16).

Only the promised purchase of an as-yet-incomplete product constitutes a government shelter. State support of the industry includes financing, regulation, and tariff and non-tariff barriers. Nearly all states encourage aerospace exports by offering financing support to foreign purchases of domestic goods.[13] Frustration over claims of predatory financing schemes soured relations among France, the United Kingdom, the United States, and West Germany, and led to a financing accord known as 'the Commonline' in August 1981.[14] While this eased tensions, it did not apply to corporations which remained free to creatively finance aircraft sales. US aircraft manufacturers complain Airbus Industries is subsidised at so many levels of production that it can offer 'flexible and imaginative financing programs' not available to the privately owned US manufacturers (see Mann, 1981a: 25; 1981b: 16-7; 'US Officials See . . . ', 1980; Washington Roundup: Financing Flap', 1980; Washington Roundup: Predatory Financing', 1980; and 'US Officials . . . ', 1981). This remains a difficult issue, the resolution of which may require GATT-level negotiations.

Though often viewed by industry personnel as a costly nuisance, government regulation can support the industry. Compliance costs 'keep fly-by-night, shoestring operators out of the marketplace' (Bluestone *et al.*, 1981: 166; Caney, 1982: personal correspondence). Sometimes certain industry segments are positively affected at the expense of others. The 1985 emissions and noise standards rendered industry workhorses such as Boeing's 707s and 727s obsolete unless retrofitted with new generation equipment. This boosted sales of aerospace manufacturers, while adding to the financial burdens of airlines. At times standards implementation has been relaxed to promote the export of domestic products. Following the 1979 DC10 crash at O'Hare, the Airline Passenger Association, Inc. sought a judicial order to ground all DC10s until it was determined that a structural flaw had not caused the accident. Then FAA Administrator Langhorne Bond responded he 'had a duty to promote aviation commerce and therefore couldn't act solely on the basis of safety and accede to the plaintiff's demand' (United States House of Representatives, 1980c: 93; see Chapter 4 for more detail).

Finally, states support the industry by erecting non-tariff barriers. While these violate the intent and often the letter of various bilateral and multilateral agreements, they are also difficult to identify or prove, making them a popular means of support. They are also compelling because they 'lie within the power of an individual nation' to implement (Cowhey and Long, 1983: 162). The NTBs of greatest interest to this study are standardisation and safety requirements. (For a discussion of other kinds of industry NTBs see Appendix 2.) Industry personnel are concerned that safety regulations and engineering standards may be used to preclude foreign products from markets even though the agreement on technical barriers to trade negotiated under the auspices of the General Agreement on Tariffs and Trade prohibits their use as 'obstacles to international trade' (Lowndes, 1982: 25).[15]

INDUSTRY DYNAMICS

A Risky Business

Civil aviation's dynamics reveal its high risk nature. They simultaneously compel states to intervene to facilitate participation in a strategically valuable industry and induce corporations to pursue joint production strategies to reduce risk and gain market access. The 1970s emergence of surplus capacity in both the service and manufacturing sectors of civil aviation intensified the effects of enduring industry characteristics. For 'a sustained period of time and a large percentage of all producers, demand [was] not sufficient to absorb enough output for prices to sustain substantial employment and adequate returns on investment . . . ' (Strange, 1979: 304).

Susan Strange and others have identified a number of causes of surplus capacity, including the presence of new and effective competition, skyrocketing cost of investment to produce a unit of output, and a changed pattern of demand (see Strange, 1979; Cowhey and Long, 1983; Tsoukalis and Ferreira, 1980). The presence and proliferation of surplus capacity in several industry sectors was not anticipated by state or corporate elites. This made policy selection and implementation difficult, especially when combined with the conflicting demands within the civil aviation sector.

At the systemic level, changes in the international market structure have been tremendous: production and sales strategies must be globally conceived. No aircraft manufacturer can profitably produce for a domestic market only. Realising this, non-producer states can generally bargain for some level of compensation in return for purchasing a particular aircraft. In addition, other producer, as well as non-producer, states have gained in relative power capabilities since the Second World War. They have learned to negotiate more effectively and to compete in the market place. As the market approaches a unified global structure, the anomaly identified by Deputy Secretary Shane at the outset of this chapter becomes more apparent: separate nation states operating competitively within that unified market does not make sense.

As Harry Johnson pointed out in 1970:

> In an important sense, the fundamental problem of the future is the conflict between the political forces of nationalism and the economic forces pressing for world economic integration . . . in the longer run economic forces are likely to predominate over politics, and may indeed come to do so before the end of this decade. (Gilpin, 1975: 220)

Two other layers of contradictory demands added to the confusion of elites as they attempted to select and implement appropriate and effective domestic and international policies. The first was a dramatic shift in domestic preferences. The second was the emergence of an increasing number of issue area linkages not directly related to aviation.

Industry Characteristics and Surplus Capacity Variables

Competition among commercial aircraft manufacturers in the world is intense. Commercial airliner market dominance is not permanent: In the 1930s and 1940s, Douglas was the leader with its DC3, but by the late 1950s, Boeing's 707 had gained pre-eminence. Hope springs eternal for the manufacturer who can offer the right product at the right time.

Airline competition is also intense, especially since economic deregulation. Prior to deregulation 36 commercial airlines were certificated and engaged in scheduled domestic air service. Between 1978 and 1985, 197 carriers came in and out of the industry. By the end of fiscal year 1986, 46 commercial airlines remained engaged in scheduled domestic air service. Until recently most non-US airlines enjoyed monopoly status as a result of state policy.

Market success promises benefits ranging from increased profitability to the ability to influence manufacturer decisions regarding new generation aircraft specifications. The costs of failure include acquisition, bankruptcy or liquidation. Airline success is also ephemeral. The original 'Big Four' US airlines were American, Eastern, Trans World, and United airlines. This changed in 1984, when American, United, Delta, and USAir earned 60 per cent of industry operating profits and 102 per cent of net profit. It changed again in 1985, when American, Delta, and Eastern earned operating profits of $1.2 billion, or 60 per cent of total industry profits. Then, in 1986, the 'new' Big Four – American, Texas Air Corporation,[16] Delta, and United – accounted for 60 per cent of total industry traffic, nearly six percentage points higher than the combined results of the original group. The story of Eastern Airlines exemplifies the industry's volatile profitability records. In 1985, it was one of three airlines earning 60 percent of industry profits; in 1986, it nearly declared bankruptcy and its dire financial straits convinced Department of Transportation officials that acquisition by Texas Air Corporation was necessary.

Manufacturer success requires satisfaction of airline performance, maintainability, and efficiency needs and government regulations.

Commercial class aircraft are expensive to produce. McDonnell Douglas spent $2 billion on the DC10. At a cost of about $40 million per plane, 400 DC10s had to be sold for its costs to be amortized, and for the corporation to realise profits (Smith, 1979: 63).

Today, airframe development requires $4–5 billion and 4–6 years, and a new engine $1.5-2 billion and 6 years (*The Competitive* . . . , 1985: 5). The return on a successful investment takes 10–15 years. It could take longer if sales are slow, the market requires a derivative model, productivity does not advance as rapidly as anticipated, or interest rates rise (see Figures 3.4 and 3.5).

Figure 3.4 Launching cost – new large commercial transport aircraft (one model – airframe only – constant dollars)

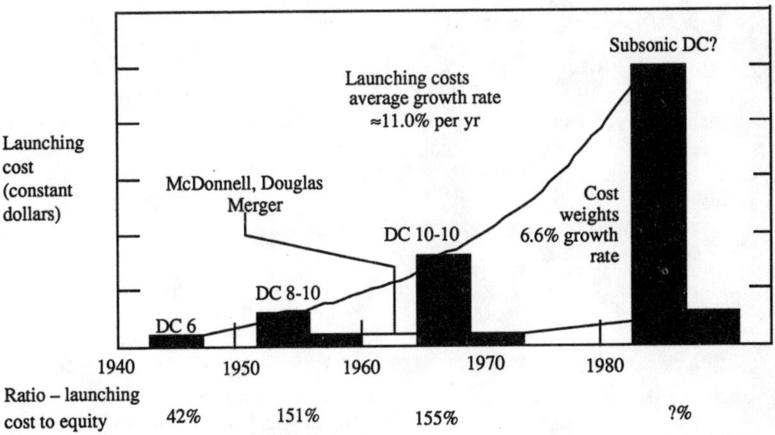

Source: Douglas Aircraft Company; Taneja (1980: 146).

Similarly, replacing a fleet of aircraft is expensive and requires advance planning. In 1986, the average commercial transport cost $31 million and the average fleet of a US domestic airline had 115 planes. Approximately 100 needed to be replaced representing a capital investment of over $3 billion (*AIA Aerospace Facts and Figures, 1986/87*, 1986: 26; *Air Transport 1987*, 1987: 8–9). Intense competition has drained airline cash reserves when ageing fleets, government regulations, consumer needs, and fuel costs are dictating the arrival of a third 're-equipment cycle'.[17]

Airlines must anticipate their needs by at least ten years if they are to have any significant participation in the design and technologies of new generation equipment. Unfortunately, it is

Figure 3.5 Cumulative cash flow for medium-size aircraft

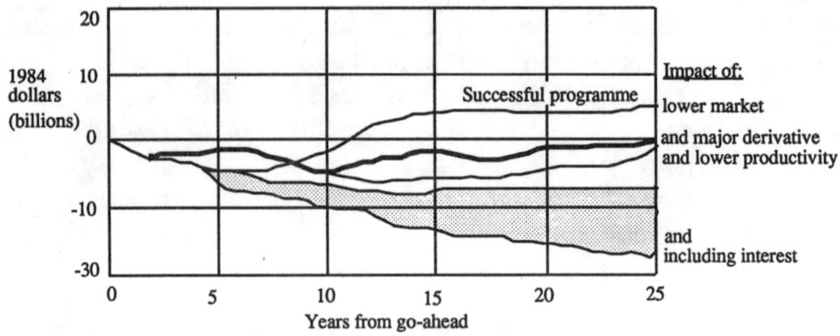

Basic curve adopted from report of Aviation Advisory Commission

Note:
Assumptions for'successful programme' line include 1) 700 units sold during the first 10 years of deliveries following development, with a total of about 950 aircraft delivered over the 25-year period shown; 2) 4-year development period; 3) customary advance payments by airlines; 4) production costs at typical US standards. Assumptions for the 'lower market' curve include 1) 280 units delivered over the first 10 years after development; 2) 5-year development period; 3) a total of 450 aircraft delivered over a 25-year period.

Source: Bacher, 1984: 14.

difficult to predict market needs beyond a 5-year time frame[18] because of the difficulties associated with trying to predict national, regional and global economic vicissitudes. Airlines must be cautious when deciding when and which aircraft to purchase. They cannot afford to be burdened with aircraft that do not meet their needs.

Airlines need aircraft deliveries to be on time to minimise operations costs and to coordinate aircraft usage with schedules planned months in advance. They operate on slim profit margins and it is important they be able to take advantage of any potential new gains in operational economies.[19] They want to be the first to utilise new equipment because of the efficiencies and prestige that accompany the privilege. They view using new equipment as an important selling technique and advertise the acquisition and use of the latest generation.

This helps explain why market timing is so critical. Although the entire development process can take as long as ten years to complete, delivery on time can mean the difference between successful market penetration or a white elephant. As little as a six-month delay in introducing new generation equipment can cost a manufacturer 60 per cent or more of the total market; '70 per cent of the market for a new aircraft entry is exhausted in 18 months' (Bluestone *et al.*, 1981: 8, 81).

For example, the Douglas DC8 was ready for airline use less than a year after Boeing's 707, but that delay allowed Boeing to capture the lion's share of the original commercial jet market, a position it has maintained ever since. In 1970, the DC10 preceded Lockheed's L1011 to the market by only a few months, but this was enough to cause Lockheed's unit sales to fall below the breakeven point. Eventually Lockheed withdrew from commercial aircraft production. Manufacturers who deliver earliest and with the strongest assurance there will be no delay in contract performance gain another significant advantage in the marketing game: they have more time to capture international markets. Typically, aircraft are not marketed to foreign airlines until after they are available to the domestic market. Whichever aircraft enters the market first has the advantage of building a broad support base for external sales. Airlines using the aircraft lobby foreign airlines, whose airport destinations they share, to purchase the same aircraft and lower long-term operational costs via maintenance sharing agreements. Also the manufacturer will be able to make performance, payload and cost promises on the basis of actual figures. The sales strategy is to gain one key customer in the area. In 1972, McDonnell Douglas and Lockheed both courted Turkish Airlines (THY) as a key customer, believing the aircraft it selected would be the choice of most Middle Eastern airlines.

Critical Nature of International Markets

International markets have always been taken seriously by aircraft manufacturers. Export sales generate substantial revenues and help offset the high costs and risks of aerospace development programmes. They provide the direct benefit of increasing profitability by selling more aircraft. They also provide two indirect benefits: product support sales and lowered cost of unit production. Figure 3.6 demonstrates the long-term income generated by the spare parts, personnel training and maintenance packages associated with aircraft sales. Figure 3.7 reveals how progression along the productivity learning curve can have a profound affect on economies of scale.

The global market is more important today than ever. Since

Figure 3.6 Deliveries of US commercial jets, spares and modification work

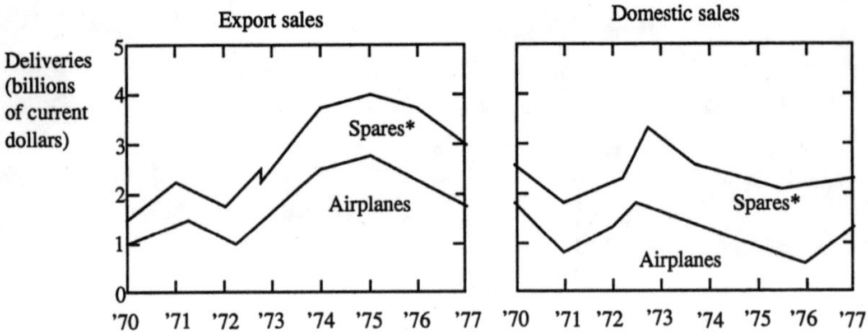

* Includes initial spares, follow-on spares, and parts and materials used in modification work (e.g., converting older passenger jets to cargo airplanes).

Source: Boeing Commercial Airline Company; Taneja (1980: 143).

Figure 3.7 World market importance – aircraft production costs

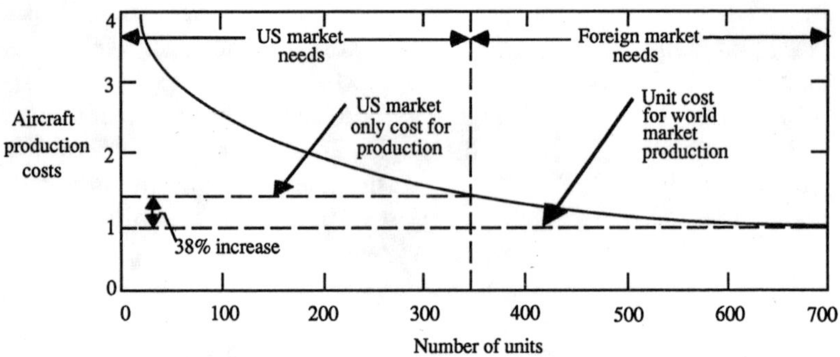

Source: Douglas Aircraft Company; Taneja (1980: 145).

ninefold to $2 trillion (Culley, 1987: 1). International trade in civil aviation goods and services has contributed substantially to this trend. United Nations figures[20] reveal that total free world exports of civil aircraft,[21] engines and parts were $25 billion in 1985 (Yager, 1987: 3); air transport as measured in revenue passenger miles (rpms)[22] has grown 7 per cent in the last decade ('Continued International. . . .', 1987: 5).

Figure 3.8 reflects changes in the importance of international markets. Immediately following the Second World War, US aircraft commanded 85 per cent of the world's markets. Today competitive foreign aircraft are sold in their 'home' markets, which have increased in size due to the internationalisation of production, and have begun to penetrate former American strong holds. Figure 3.9 and Table 3.1 reveal changes in the structure of the world's passenger traffic market. By the 1990s, the US share will have decreased to 36 per cent from 55 percent in 1970, while market segments in the Middle East, Latin America, the Pacific, and Asia are expected to increase from 16 per cent in 1970 to 33 per cent.

The increased importance of international markets affects aircraft manufacturers in three ways. First, as technological advances become more consistently available in products, initial price and financing of aircraft become more important to their sale. If companies are responsible for financing aircraft sales, then requirements for raising capital increase along with financial exposure to risk; competition with foreign governments in the negotiations regarding sales and financing is virtually guaranteed.[23]

Second, most non-US airlines are government-owned or supported and aircraft purchases are typically a political issue. Negotiations nearly always include at least one government (*The Competitive* . . . , 1985: 3, 51-6; Gidwitz, 1980; Lawler, 1985; Willy, 1986; Bracken, 1986). As Elaine Bendell, public relations officer for

Figure 3.8 Market share of world aircraft and parts exports, 1975–85

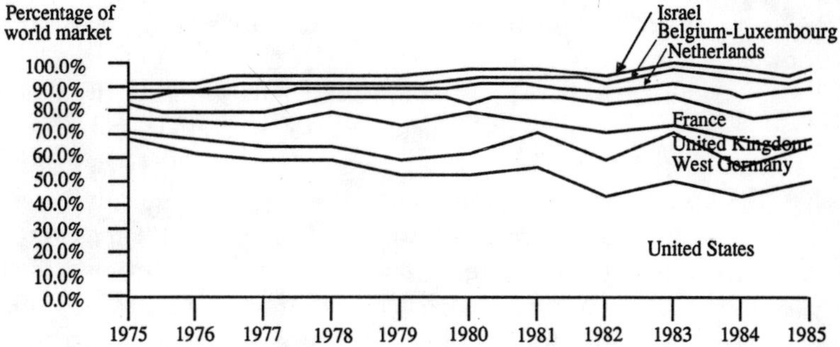

Source: United Nations, *1986 International Trade Statistics Yearbook*, 1988 pp. 236, 753–4, 1238–9.

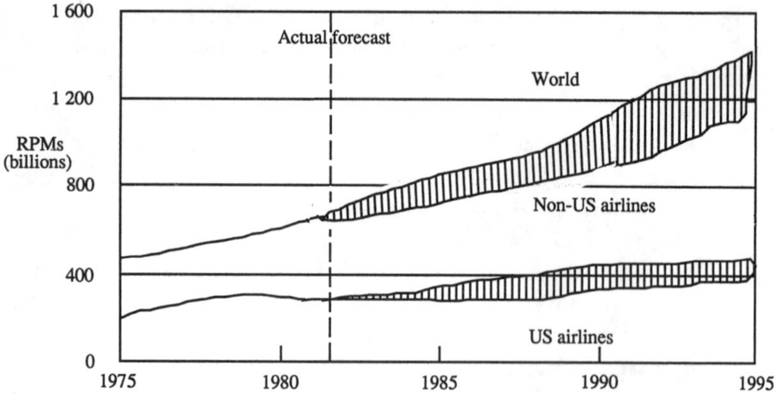

Note: Excludes USSR and non-ICAO nations, but includes Taiwan.

Source: The Boeing Commercial Airplane Company.

McDonnell Douglas, explained, manufacturers are forever mindful of 'the three "Ps": Price, Performance, and Politics' (1983: personal interview). Because of civil aviation's strategic value, more states are bargaining for participation in the production process in exchange for purchasing the aircraft. Third, the corporate policy of building aircraft to meet US carrier requirements, then modifying them for international market needs may be reversed (Lawler, 1985). If air traffic increases in the Third World as dramatically as has been predicted, the airlines which dominate these regional markets will have more to say about aircraft specifications. In addition, their state elites will be in a position to bargain for substantial offset or co-production agreements as part of the aircraft acquisition contract.

US airline responses to economic deregulation have weakened their ability to act as launch customers for aircraft manufacturers, a role played since the end of the Second World War.[24] First, route structures were shifted to a hub-and-spoke network requiring a return to smaller aircraft. Airlines with such planes will use them as long as they meet noise pollution requirements,[25] fuel prices remain low[26] and labour costs continue to drop. Since 'routes and traffic cannot be forecast credibly for the next 12 months', airline management will postpone aircraft purchasing decisions; it is too difficult to calculate the investment return (*The Competitive . . .* , 1985: 36). Second, airlines initiated fare wars to capture new markets which

Table 3.1 Projected growth in air travel

	Share of traffic (per cent)		
	1970	1982	1995
US	55	40	36
Europe and Canada	29	31	31
Rest of world	16	29	33
	100	100	100
Revenue–passenger miles (billions)	288	678	1,413

Source: The Boeing Commercial Airplane Company

decreased profitability: in the four year period between 1980 and 1983, US certificated air carriers incurred operating losses totalling almost $1.6 billion. In 1986, the industry posted a net loss of $223 million (US Department of Transportation, 1987: 19-46). Third, airlines rationalised production via mergers and acquisitions, and in so doing, obtained aircraft. These responses significantly reduced airline demand for aircraft.

Domestic Preference Function and Issue Area Linkages

The effect of these airline responses was exacerbated by the existence of a surplus pool of aircraft available at 'rock bottom prices' (*The Competitive* . . . , 1985: 35) and the coincidence of the 1980-1 recession with the initial deregulation phase, which further reduced passenger demand and airline profitability. Skyrocketing fuel prices prolonged the recession and wreaked havoc with operating cost formulas used by airline management. Ironically, by the time fuel-efficient equipment was available, oil prices had fallen dramatically. Because new generation equipment is designed to fit perceived future market and government requirements, a problem arises when those once perceived needs become irrelevant or even a liability. In the 1960s, jumbo jets with vastly increased capacity were introduced to accommodate a predicted dramatic growth in air travel. However, the combination of three wide body jets and elastic consumer demand led to a problem of surplus capacity and airlines lost money on their new, 'over efficient' investments. More recently, Boeing, McDonnell Douglas and Airbus aircraft were designed to fit changed government and market requirements, and were available when airline fleets were ageing and needed to be replaced. Anomalously, airlines could not afford the new equipment (Welling, 1982: 65).

Boeing's experience was typical. Airlines placed a lot of orders in 1980-81. Then sales stagnated; there was a nine-month hiatus between 767 orders. Next, orders were delayed or cancelled: by June 1982, United Air Lines planned cancellation of twenty 767s valued at $40 million each, and American Airlines cancelled its order of fifteen 757s for a total of $600 million. Air Canada and British Airways delayed delivery of 12 and 19 767s respectively representing a value of over $1 billion until low interest Eximbank loans were forthcoming.

During the 1970s, state decision makers felt heavy pressure from domestic constituencies to change previous policy patterns. In the US, a declining economy coincided with the end of the Vietnam conflict. Environmentalists and defence critics shared interests regarding the aerospace industry, rendering their lobby efforts effective. Together, they severed all federal funds for a US supersonic transport, stopped the B-70 manned bomber programme, and dramatically reduced the B-1 bomber development funding.

The industry's need for radically higher research and development funding corresponded with the simultaneous collapse of military and commercial markets. Post-Vietnam military de-escalation significantly reduced the defence budget. Between 1968 and 1971, industry sales to the Department of Defence dropped 24 per cent (*AIA Aerospace Facts and Figures 78/79*, 1978: 81). At the same time, airlines were suffering from overly optimistic purchases of jumbo jets; during the identical 1968–71 time period jet transport sales dropped 68 per cent (*AIA Aerospace Facts and Figures 78/79*, 1978: 81). A bust cycle followed the mid-1960s civil aviation boom.

European governments responded differently. State elites recognised the potential benefits inherent in dominating the industry and controlling the safety regime. Between 1967 and 1971 US state support increased by 16 per cent, while France increased its by 27 per cent, West Germany by 82 per cent, Great Britain by 21 per cent, Japan by 147 per cent, and the Soviet Union by 61 per cent (United States House of Representatives, 1976: 81–5).

By the early 1960s, European governments recognised they would have to collaborate to compete successfully against the US. The British and French governments financed the development of the supersonic Concorde via a bilaterally organised consortium. Hoping to challenge US pre-eminence in aircraft technology, they continued the project even after it became apparent that short- (and long-) term profits were not likely.

> European leaders were apparently willing to suffer substantial economic losses if domestic manufacturing capacity could be re-incubated, and if their own national firms could begin to produce import substitutes for the large number of US-built aircraft being purchased (Bluestone *et al.*, 1981: 46–47).

The vast amounts of investment capital needed could not be generated privately, not even by a single nation. The stakes involved in the gamble were too high. Thus began the move to multinational projects wherein profits and losses could be shared.

The lessons learned regarding management, financing, production and marketing on a shared basis were transferred to other aircraft project consortia, including the largest and most successful to date, Airbus Industries. Begun as a French-German attempt to build a two-engine, medium-range, wide-body transport capable of competing against American products, Airbus Industries eventually sought other members to help spread risk. Today's consortium has six members.[27] Evidence of the consortium's success is found in the fact that in 1986 Boeing and McDonnell Douglas won 65 per cent of world orders for commercial passenger planes, down from 85 per cent immediately following the Second World War, and 80 per cent in 1985 (Rudolph, 1987: 50; Comes, 1987: 80–1; 'The Technology Edge', 1987: 1). Airbus' original goal was to capture 30 percent of the world market (Lawler, 1985).

Internationalisation of Production

Both the US and Europe have adopted internationalisation strategies they hope will facilitate their adjustment to structural changes in the environment, and may even alter it in such a way as to be more beneficial to them (see Axelrod and Keohane, 1985: 228; Gilpin, 1981; Zysman and Tyson, 1983; Shepherd, Duchene and

Saunders, 1983). Internationalisation is not a new concept. In the 1930s Fokker built DC2s in Europe under licence; the Soviet Union produced the DC3/C47 as the Li-2; Peru co-operated in building the Faucett-Stinson aircraft. Still, the false starts at co-operation far outnumbered the successes until the 1970s (Lawler, 1985: 13). In today's world, benefits believed to accompany this response outweigh perceived costs.

First, international co-operation effectively reduces RD&P risks. It encourages the pooling of resources and talent that facilitates a more efficient production process. No single company or country is forced to bear the incredible costs associated with market failure, and each gains benefits equal to its initial investment. In spite of the overall trend of government support for civil aviation RD&P efforts, the temporary reduction of US government funding in the late 1960s and early 1970s provided a strong incentive for the growth of international joint ventures (Bluestone *el al.*, 1981: 160).

Second, internationalisation creates an opportunity for specialisation in a product or through product support such as marketing, repair and modification work. Suppliers of specialised goods and services can market to prime manufacturers in different countries and achieve economies of scale associated with increased sales. This drives down the cost of aerospace products, directly benefiting aircraft producers, and indirectly benefiting airlines and their consumers, the travelling public.

Third, the domestic markets of most states are not large enough to sustain sufficient production runs capable of supporting the manufacture of commercial class aircraft. International production increases the size of the potential home market.[28]

Fourth, international collaboration can be an effective way of gaining access to technology. Although this is a controversial issue, the general consensus is that the technology which is transferred will not undermine the competitive position of participants involved in co-production projects. There is a greater concern that denying a potential foreign partner's access to a production market may create deep-seated resentments propelling it to become an independent competitor in the future (Golich, 1987).[29]

Fifth, cooperative efforts reduce the likelihood of producing directly competitive aircraft, such as the L1011 and DC10, thereby increasing an aircraft's chance of survival in the marketplace. For example, should each of the firms in the Airbus consortium build its own aircraft to compete with the B757, more than one would fail because the market is not large enough; by producing only one competitive aircraft, the chances of success are greater.

Sixth, and perhaps most important for its potential affect on global relations writ large, is the evolution of a new skill in managing transnational production, especially in a strategically valuable industry, among partner firms. Anthony J. Lawler, Marketing Director for Airbus Industries of North America, argues that the successful management of the Airbus consortium, is in part, the product of lessons learned from the Concorde project (1985; see also *The Competitive . . .* , 1985: 63–4).

The agreement between the US's General Electric Aircraft Engine Group and France's *Société Nationale d'Etude et de Construction de Moteurs d'Aviation* (SNECMA) in the development of the CFM56 commercial aircraft engine provides an example of the benefits possible with this kind of arrangement ('US Lifts . . . ', 1973: 37). The project involved the transfer of B-1 bomber engine technology from GE to SNECMA counterbalanced by two factors: 1) SNECMA contributed half of the CFM56 development capital, which could not be internally generated, nor extracted from the US government; 2) A virtually guaranteed market share for the final product, stemming from the relationship between SNECMA and the French government and the latter with Air France, increased the chances of market success for the final product. While some technology, jobs and profits were transferred outside the country, a substantial amount of the same benefits were reserved for the US. The CFM56 engine has been tremendously profitable. Corporate and state participants in this joint venture have reaped many benefits from its success.

SUMMARY AND CONCLUSION

The preceding discussion showed why states perceive aerospace as strategic and therefore wish to participate substantially in the industry: it is vital to a state's pursuit of many political-economic domestic and foreign policy strategies and goals. Economically, the industry provides employment, generates revenues and maintains a position in high technology. These translate into political benefits for states: economic benefits add to (or detract from) the health of a domestic economy, which helps states satisfy domestic constituencies and legitimately hold on to the authority to allocate values in society. A strong domestic economy also strengthens a state in its international bargaining position. In addition, aviation is a vital element in maintaining national security and can be used as a political instrument. That states perceive the industry to be a vital asset is confirmed by the various supports and encouragements given the industry.

This analysis also detailed the risky business dynamics of both the manufacturing and service sectors of civil aviation, and demonstrated how elements of surplus capacity, an altered market structure, changes (even if temporary) in the domestic preference function, and issue area linkages combined to bring changes in the regime. Aircraft manufacturers struggled to meet global market and state requirements in the context of a radical increase in production costs. In the 1970s they lost significant state support as American leaders attempted to satisfy a perceived constituency mandate to spend less on aerospace and more on social benefits. Subsequently manufacturers lost consumers as airlines struggled to survive in their own radically changed political-economic structure. As research and development monies dried up and European competition intensified, US manufacturers sought international partners to provide needed capital and virtually guaranteed markets.

Europeans were motivated to co-operate so they could compete successfully with US products. Collaboration created a larger 'domestic' market and more research, development and production funds than could be generated by a single state. These co-operative efforts represent much less of a philosophical departure for European governments than for the US. As noted in Chapter 2, Europeans are comfortable with a 'participatory enterprise' approach to state-economy relations, while Americans prefer a 'regulated enterprise' approach, keeping the state and the economy as separate as possible. The result of all these corporate responses may be international corporatism (Strange, 1979).

In air transport, state responses to surplus capacity are triggering production rationalisation and may lead to internationalisation as well. American policy makers believed economic deregulation of air transport services would benefit consumer and operator alike as routes and fares were dictated by the marketplace. The rest of the world viewed US behaviour as an attempt to secure perpetual hegemony by playing a 'survival of the fattest' game which the US was certain to win.

Initially the Europeans and others balked. But, by mid-1984, the UK and the Netherlands began their own domestic and bilateral economic deregulation movement. The rhetoric is similar to that in the US, but their motivations could also be less philosophically pure. Perhaps Europeans are hoping to take advantage of the current weakened position of American carriers as they adjust to the final throes of the rationalisation process. European national carriers rationalised long ago so this might be a good way to grab markets and usurp a dominant position in this civil aviation sector. Given the transnational management lessons learned by Europeans in the production of the Concorde and the Airbus family of aircraft, it may be a fairly short step to the creation of a few transnational mega-carriers.

States can assert influence to the extent they increase their power resource bases and their ability to translate them into desired outcomes. As power resources are more equitably distributed and negotiating skills sharpen, other states may attempt to restructure a regime to make it more beneficial to them. Alternatively, they may prefer using circumventing power within the regime to reap an increasing number of benefits (Aronson, 1977: 17–19). Difficult political or economic problems impel governments to be more assertive in their pursuit of regime dominance.

This analysis of civil aviation's value and dynamics has also shown how complacent, nonfeasant regime management can lead to internal unanticipated changes. State elites operate under conditions of complex interdependence in which issue area linkages are often not mutually supportive, but rather have conflicting goals and strategies. Domestically the same state elites must respond to constituency demands which are increasingly marked by deep cleavages regarding appropriate or desirable policy choices. Under pressure from both domestic and international sectors, decision makers momentarily concentrate on some issues to the neglect of others. Given the evolving market structure, production technology and management skills, this was enough to allow corporations to pursue new RD&P strategies. Corporate elites make decisions in somewhat less complicated surroundings. Their goals are more focused; they seek corporate survival. Their responses to changing international and domestic structures may create new constraints on state behaviour, leaving state elites with fewer policy choices to satisfy a widening array of demands.

Finally, the foregoing demonstrates the crucial role of process in effecting change. It is the daily operation of states and corporations which leads to change in today's world. Seemingly insignificant everyday decisions can trigger significant change if the political economic conditions in place are vulnerable.

Disaster can ensue when safety regulations are neglected. The potentially dangerous combination of industry value and dynamics coupled with perceived threats to the existing distribution of power in the international system can provide incentives for negligent, nonfeasant behaviour by both state and corporate actors.

Chapter 4 provides a detailed analysis of two catastrophic crashes with tragic consequences that exemplify some of the failures of the aviation safety regime and help identify the loopholes which allow the intrusion of politics and economics into the resolution of what should be purely technological issues. Each reflects various aspects of industry dynamics which combined with political-economic considerations and the existing certification process to produce deadly results.

4 Disaster in the Skies

> We were cast and wrought and hammered to design, . . .
> But remember, please, the Law by which we live,
> We are not built to comprehend a lie, . . .
> . . . , for all our power weight and size,
> We are nothing more than children of your brain!
> (Rudyard Kipling, 'The Secret of Machines')

Commercial aviation, vital to global transportation and communication, must function reliably and efficiently to facilitate the 'freer trade' regime upon which the world is increasingly dependent. Chapter 2 disclosed how the fledgling industry was unable to regulate itself. Too many pressures and incentives to perform beyond safe parameters existed even though safety was considered a significant selling point to encourage a timid public to fly. State regulation was required to achieve acceptable levels of safety domestically. Internationally, civil aviation regimes – transport, trade, safety – were created to facilitate aviation operations. These endured forty years, as long as they were perceived to be effectively and fairly managed. Perceptions of fairness vary. What may be considered fair in one era, based on power asymmetry in the international system, may be considered inequitable in an era characterised by more equal power resource distribution.

International civil aviation regimes resemble those characterised by hegemonic stability – they were created and maintained by a hegemonic state and willing state and corporate participants.[1] The hegemon was primarily responsible for regime maintenance. Its effectiveness rested, in part, on the presence of a power resource base including the leadership ability to translate resources into desirable outcome. Other factors included domestic support, reinforcing and supportive issue area linkages and participant agreement regarding the need for and means of issue area management.

Once in place, the value of regimes in facilitating efficient aviation operations meant that, in spite of environmental changes which rendered the hegemon less able to maintain the regime, the result was internal change rather than complete collapse.[2] Management rules and decision-making procedures changed as global political and economic structures and conditions evolved,[3] caused in part by advances in technology and knowledge.

A combination of factors can undermine hegemonic leadership and possibly initiate change. First, a redistribution of power and wealth among participants can alter the structure of the international management regime. This can happen when the hegemon's relative power position, in terms of both possessional and relational power, is weakened.[4] A part of this phenomenon may be advancement along a learning curve by participant states regarding how to bargain and manage, as well as how to build and fly airplanes, more effectively. For example, changes in Europe's approach to safety regulation were triggered by dissatisfaction with the US handling of the 1974 and 1979 DC10 disasters. Changes were possible because of Europe's renewed ability to compete in the commercial class aircraft market and the transnational management lessons learned along the way as discussed in Chapter 3.

Second, significant changes in domestic and international environments can require state and corporate adjustment. Chapters 2 and 3 provided evidence of eroding US domestic support for a state role in aerospace. By the late 1960s, Americans were disenchanted with government spending on the military-industrial complex and growing weary of the costs associated with US leadership in several international regimes. Regime participants were increasingly dissatisfied with American dominance. While still convinced international commercial air transportation needed management, Europe was less pleased with its manner of management. They demanded changes in the governing Bilateral Air Transport Agreements, reinvigorated their commercial class aircraft industry, and began developing their own safety regulations.

Third, conflictual rather than complementary issue area linkages can render policy decisions ineffectual and contradictory. The emergence of a potentially undermining complex interdependence in recent years is reflected in Chapter 3's discussion of civil aviation's strategic value in the twentieth century: it must be safe to continue to be useful as a tool of political and economic, foreign and domestic policy, but individual policy goals may oppose each other. Chapter 4 demonstrates how research, development, production, marketing and maintenance processes and costs can conflict with regulation creation, product certification and maintenance monitoring processes and costs. This helps explain why US state elites, concerned about boosting the domestic economy, supported the 'All-American' DC10 over the L1011 which used English Rolls Royce engines. It also suggests some regulatory nonfeasance, prompted by domestic political and economic concerns, which led to a poorly designed, constructed and maintained DC10. In the long run, US credibility as an authority capable of maintaining a satisfactory safety regime was diminished.

Fourth, procedural rigidity can prevent appropriate adjustment to technology and knowledge advances. Chapter 4 indicates how inflexible bureaucratic and process structures reduced FAA competency, effectiveness and safety function performance.

Fifth, complacency, characterised by an 'if it works, don't fix it!' attitude, can thwart awareness of environmental changes necessitating management responses. Complacency can be associated with a reputation for conservative and safety-oriented manufacture and with perceptions of aircraft and air transport being safe. In the presence of an adequate aviation safety regime, participants may not risk the introduction of potentially costly changes.

Therefore, sixth, crises may be the necessary catalysts for change. Changes in civil aviation regulations often await some sort of crisis.[5] Accident related modifications may be implemented within six months, whereas technology induced changes may require 4 to 6 years for implementation (Chapman, 1981). The case studies herein stress the role crises play in regime change,[6] providing evidence they alert state elites to altered circumstances. So many other issues need attention that those less urgent are left alone until they reach crisis proportions – the 'fire fighting' concept discussed by Cowhey and Long (1983: 165).

Maintaining an effective aviation safety regime is not easy. State and corporate actors still confront tremendous political and economic pressures to minimise costs, to introduce new technology, and to deliver aircraft and the travelling public to their respective destinations on time.

> Pressures to prevent effective . . . regulation will always persist so long as the profits from regulation avoidance exceed the probable cost of compliance and so long as corporations have a fiduciary duty to their share holders to maximise revenues. (United States House of Representatives, 1980c: 5)

Chapter 4 reveals what can happen when regime rules fail. At the most visible level are the DC10 catastrophes and the tremendous losses associated with them. More significant were the longer-term effects: the crashes and myopic responses by US state and corporate elites helped confirm a European perception that only co-operation could enable them to gain autonomy from American hegemony.

REGULATORY PROCESSES

An aviation safety regime is founded on three major processes: regulation creation, component certification, and maintenance and operations oversight. All three are fraught with political, economic and institutional interference.

Regulation Creation

Federal Aviation Regulations (FARs) govern virtually every aspect of safety in civil aviation.[7] Their creation is an exercise in compromise, which attempts to balance technological capacity and cost burden at three levels of safety. Most basic are inviolable principles which allow no room for compromise, such as the provision of seat belts.

Next are marginal safety measures, difficult to justify on the basis of a cost-benefit formula. For example, aircraft interiors cause unnecessary deaths in otherwise survivable crashes.[8] It is virtually impossible to calculate whether the number of survivable crashes and the potential extra lives saved outweigh the additional costs of safer interiors.

Finally, there are 'imponderables':[9] the situation which arises when 'you can calculate what safety measures will cost you, but you can't measure how much safety you are buying for it' (Johnston, 1976: 119). These exist for all aircraft. They are multiplied a hundredfold when innovative technology is incorporated into new generation aircraft and engines. Testimony to the incredible strain imponderables place on a safety regime is found in the fact that 'no new aircraft before the 747 flew as much as half a million hours without a fatal accident' (Eddy *et al.*, 1976: 33).

Because regulation creation is an inexact science, an attitude accepting of compromise permeates the entire process. Thus the potential danger of compromising even inviolable principles can become real. Contributing to this attitude is the fact that, like any applied science, designing and building an aircraft is an exercise in compromise.

Complacency can exacerbate the effects of an attitude which accepts compromise as a 'necessary evil'. Although a 1980 National Academy of Sciences study noted 'aircraft safety demands a "forgiving" design that is tolerant of failure' (Low, 1980: 3–4), design decisions 'on sophisticated systems are being predicated daily on assumptions that certain in-service maintenance programs will be applied, and [that] operator's maintenance programs are being approved, based on assumptions that design considerations provide certain levels of performance after element failure or malfunction' (United States House of Representatives, 1980c: 97).

Certification

Certification is the process which determines whether or not an aeronautical product conforms to established standards. It involves three phases – design, manufacture and maintenance. Its goal is to insure that aircraft used by airlines are designed and assembled to be safe and then are maintained in an airworthy condition. 'The concept of "airworthiness" must be as unambiguous as possible. The highest level of safety . . . cannot be achieved with even slightly defective aircraft' (United States House of Representatives, 1980c: 6). The process is not always impeccably and disinterestedly executed (see Appendix 4).[10] Certification must consider 'the interrelationship between aircraft design, manufacture and maintenance' (United States House of Representatives, 1980c: 1).

Maintenance and Operations Oversight

Maintenance and operations oversight should protect an aircraft's airworthiness status. Maintenance flaws are dangerous because they can ruin aircraft airworthiness. Nevertheless, the incentive structure, especially after economic deregulation, dictated that all operations-related costs be cut (Golich, 1988; O'Brien, 1987). Arvin O. Basnight, former director of the FAA Western Regional Office, explains:

> Maintenance is one of the easiest things to stretch. You can't

> stretch a pilot's time too much, you can't stretch a fuel load too much, but you can stretch maintenance. If you can find shortcuts on how to do maintenance, you look for those. It isn't always bad. It's part of the evolution of change. But . . . as these systems are tested, we need to have the integrity of them constantly protected . . . any shortcuts in aircraft maintenance by airlines should be subject to the closest FAA scrutiny. (Dean and Kendall, 1979: 3)

Failures at all levels of the regulatory processes led to the 1974 and 1979 catastrophes. The Paris disaster was related primarily to the design phase of the certification process, the O'Hare tragedy to maintenance and operations oversight.

Process Problems

Several problems interfere with the FAA's ability to perform its regulatory functions, ranging from lack of resources to unclear communications rules. Domestic, political and economic realities make it impossible to staff the FAA adequately,[11] even though there is a recognised need 'for more qualified, critical, and analytical oversight by the senior FAA staff than is presently exercised' (Low, 1980: 21). For example, the FAA office responsible for certifying the DC10 also had to oversee the continuing airworthiness of the DC8, DC9, L1011, Rockwell International's Aero Commander, and various components of Lear-Avion and Hughes Aircraft. More than 4000 McDonnell Douglas engineers worked on the design and production of the DC10. Only 10 FAA engineers were 'intimately involved' in the aircraft's evolution (Green *et al.*, 1979: 1).

Many FAA oversight duties are delegated to manufacturers via a system of designated representatives. Designated Engineering Representatives (DERs) perform component conformity inspections. Designated Manufacturing Inspection Representatives (DMIRs) monitor production quality control. They are employed by the manufacturer. Their FAA oversight duties are secondary but their DER time is paid by the manufacturer. 'In the case of the DC10, there were 42 950 inspections. Only 11 055 were carried out by FAA personnel. The rest were done by McDonnell Douglas DERs'

(Eddy *et al.*, 1976: 180). Of 98 major fail-safe studies prepared by the manufacturer only 14 required FAA approval (United States House of Representatives, 1980c: 17).

A potential for conflict of interest associated with delegating certification oversight duties is present when manufacturers are under intense time and economic constraints.[12] 'If a DER questions a design analysis wearing an FAA hat, that questioning may require a change which is enormously expensive for the DER's employer' (United States House of Representatives, 1980c: 17). More serious, the likelihood of catching mistakes is reduced. DERs often review analyses they originally drafted. It may be unreasonable to expect them to do so with as critical an eye as is needed.[13]

As new generation aircraft, the DC10 and L1011 incorporated innovative technologies, but often were required only to meet decades-old standards.

FAA engineers are at a disadvantage because they are seldom able to maintain currency in the technological developments they are ultimately called upon to monitor.[14] There is 'greater technical competence and state of the art currency on the part of personnel in the aircraft industry than in the FAA' (Low, 1980: 5, 21). FAA engineer technical incompetence is most obvious when called upon to interpret design analyses during certification.

Inadequate numbers and quality of personnel is exacerbated by a rigid organisational structure which accounts for

> fragmenting the work of engineering specialists among many different functions, inconsistent interpretations of regulations from one regional office to another, and a lack of communication among personnel in the regional and headquarters offices on matters of common interest and experience. (Low, 1980: 22–3)

Poor communications flow during the certification process can effectively prevent FAA personnel from critically evaluating vital data. In some instances, aircraft manufacturers were able to withhold information they believed unrelated to safety. Federal Aviation Regulations provide the option to dismiss certain problems on the basis of risk analyses showing 'a failure or combination of failures is so "extremely improbable" that it doesn't have to be guarded against' (United States House of Representatives, 1980c: 20). DERs approve the assumptions, methodologies and calculations for a Failure Mode and Effect Analysis (FMEA). These are edited and abstracted into 'fault analyses' before submission to the FAA. FMEAs with undesirable findings can be eliminated.[15]

Maintenance and oversight procedures are similarly flawed. Prior to the 1979 disaster, there was no clear-cut regulation requiring the FAA be notified of changes in maintenance operations, much less that the changes be FAA approved. Nor was it clear when component problems should be communicated to other industry participants and the FAA. Langhorne Bond, then FAA Administrator, noted the Chicago accident might have been prevented if the FAA's system of defect reporting and analysis had been better.

> No pattern appeared in our computers that would have alerted us to special problems in the engine/pylon area . . . because the proper information hadn't been fed into the computers . . . because our regulations didn't require . . . that it be reported to us.(United States House of Representatives, 1980c: 4)

Finally, 'the FAA has become dangerously oriented to the needs of industry management at the expense of the travelling public' (United States House of Representatives, 1980c: 5). It can be argued that a visibly safe aviation industry is a promotion for the industry. Conversely, it can be argued that safety and economic efficiency are not always interdependent goals which can be pursued simultaneously. Contradictions inherent in trying to promote both the industry and its safety are made clear when the FAA suspends a certificate which may cost the industry millions of dollars. Direct costs of a grounded aircraft include daily traffic losses and labour costs; indirect expenses include lost market share and the associated lost investor confidence as reflected in depressed stocks. Even the loss of one-tenth of one percentage point in revenues is critical in the US airline industry where, in 1986, it amounted to about $40 million (*Air Transport 1987*, 1986: 2).

The hubris of aircraft design engineers enables them to fathom and believe in concepts necessary to develop new generation equipment. The real tragedy can be that as each participant behaves according to individual rational interests, in a manner similar to that described by Mancur Olson in *The Logic of Collective Action* (1965; see also Jönsson, 1987b: 26), this can lead to the irrational consequence of a poorly designed, constructed and modified aircraft.

THE DC10 EVOLUTION

The McDonnell Douglas DC10 and Lockheed L1011 are products of the same dialectic among manufacturer, potential customer and regulator.[16] They were inspired by Frank Kolk of American Airlines who, in 1966, decided fewer and larger aircraft were needed to accommodate the predicted 14 per cent traffic growth without increasing airway and airport congestion. Lockheed began design work in January 1967 and started taking orders in September.[17] Most aviation experts predicted the market could support only one 'airbus', as they were called. Nevertheless, Douglas began designing a similar plane in November. Top priority was given to increasing fuel and payload efficiency through weight reduction and cargo door design.

The 'Big Four' airlines – American, Eastern, TWA and United – were considered key customers. Competition to please their managements was intense. Each manufacturer committed itself to performance and payload criteria, custom-designed interiors and galleys, and cheap prices. As with any buyer-supplier arrangement, purchasing decisions were affected by history. Douglas did not have a good relationship with either American or Eastern Airlines. Relations with Eastern were considered beyond repair; it was suing Douglas for 'inexcusable delays' in delivering DC9 aircraft.[18]

Relations with American were considered salvable if the right product were offered. Incorporating American's design preferences seemed a good place to start, so its request for electric rather than hydraulic controls for fastening the cargo doors was adopted.[19] In addition to pleasing a potential customer, electric controls saved weight – 28 pounds per door and 84 pounds per plane (Johnston, 1976: 227–8).

Early in 1968, airlines began placing orders. American ordered 25 DC10s in February. In March, Eastern, TWA, Delta and a British corporation, Air Holdings Ltd, announced their orders for 144 TriStars, ' "the largest sale in aviation history" . . . for a total price of $2016 million' (Eddy *et al.*, 1976: 78).[20] It appeared the DC10 had met its demise, but, in April, United Airlines (UAL) ordered 60.[21]

The stage was set for the continued and ultimately disastrous competition for the airbus market. There were enough orders for each aircraft to justify production, and the race to the market began. McDonnell Douglas won: the DC10 rolled out of its hangar in July 1970 and first flew in September, two weeks before the L1011 rollout ceremony. Many felt Lockheed should have conceded the airbus market to McDonnell Douglas at this time. Designed for the same clientele, the two aircraft were nearly identical in looks, payload, performance, price and new technologies incorporated.[22] Lockheed eliminated weight by introducing state of the art technology; the advantages gained were lost in the additional time and costs associated with design and production. Douglas won the time race, but its 'weight reduction' programme had affected airframe integrity.

Accident Related Design Differences

Design features contributed to both accidents. The first is related to fail-safe and redundancy concepts intended to make aircraft accident proof. The former ensures that, in the case of any structural failure, other structures or systems will be able to take over. The latter refers to actual system duplication. In the early days of flying, a pilot's controls were directly attached to the surfaces he needed to manoeuvre. As aircraft increased in speed and size, the pilot required more mechanical advantage to control the plane. With the advent of jets, hydraulic systems – fluid-filled cables connected to wing and tail control surfaces – were introduced. The increased ease of handling was countered by an intrinsic vulnerability to failure, making redundancy an essential part of the fail-safe concept.[23] McDonnell Douglas was among the first to utilise triple redundant hydraulic systems.

Lockheed used four hydraulic systems in the TriStar, placing one in the cabin ceiling.[24] McDonnell Douglas installed three cables, all under the cabin floor, incorporating a successful design from its DC8s and DC9s. This saved weight and design and production costs (Eddy *et al.*, 1976: 98), but increased airframe vulnerability because a structural collapse of the cabin floor could cause a hydraulics failure.

The 'size effect' associated with the new jumbos added another layer of vulnerability.[25] Aircraft flying at altitudes where it is difficult to sustain human life must have air pressurised cabins. This can create a pressure differential of 7 to 9 pounds per square inch (psi) against the inside frame. It is difficult and costly to make internal partitions strong enough to resist such pressures, so the possibility of large holes piercing the fuselage must be 'extremely remote'. Holes caused by bird strike or engine shrapnel do occur; aside from a terrorist's bomb, aircraft doors represent the greatest potential for causing big holes. The jumbos were built according to old regulations, which mandated smaller predecessor aircraft to withstand sudden decompression caused by a hole just under 1 square foot in size.[26] The DC10 was designed to survive a hole size of 8 square feet in the front of the plane and 3 square feet in the rear. This was substantially larger than required, but smaller than the 4 square feet rear cargo door opening. Later, when asked why DC10s had not been designed to withstand the larger opening, John Brizendine, Douglas' President and then DC10 programme manager, noted, 'Doors were not considered. *Doors are supposed to stay closed!*' (Johnston, 1976: 225; emphasis hers).

The Netherlands Civil Aviation Authority (Rijksluchtvaartdienst or RLD) was concerned about 'size effect'. Hendrik Wolleswinkel recognised ' . . . aircraft were complying with and even exceeding existing regulations at the time'. But he was not convinced the jumbos were adequately protected against sudden decompression. ' "There are things you can't extrapolate from experience" '(as quoted in Johnston, 1976: 116). When KLM ordered six DC10s as a part of its consortium (KSSU)[27] order for 36, the RLD, responsible for aircraft certification, wrote to the FAA and McDonnell Douglas requesting studies and more stringent regulations. The FAA responded that 'experience so far showed no need for new requirements'. Then the RLD approached other KSSU airlines, but some, such as SAS and Swissair, could do nothing since they were contractually bound to recognise FAA criteria.

Political and economic pressures began to build: the DC10s would be delivered soon and KLM wanted the prestige of being the first in KSSU to fly it. If not certificated, the aircraft would sit at the Schipol International Airport at the cost of several thousand dollars a day. The RLD relented and certified the aircraft, but not without reservations: ' "So the RLD isn't happy with the present situation, but for this generation of aircraft, we have to live with it" ' (as quoted in Johnston, 1976: 119; see also Eddy *et al.*, 1976: 164).

A second major design difference is in the cargo door. There are four types of airframe doors. Three are safer than the fourth because their locking procedures are consequential rather than coincidental. This enhances their fail-safe qualities. McDonnell Douglas used the fourth:[28] 'over center' latches, a simple, economic and seemingly safe design. These can open under pressure, so it is more critical than ever that they be properly secured. Tragically, the DC10 back-up systems were not fail-safe. It was possible for closure to be indicated on the flight deck instrumentation when it was not complete.

Specific Certification Process Flaws

The FAA was at a distinct disadvantage when trying to determine how remote the possibility was of a DC10 door opening accidentally: they did not have all the relevant information. First, it did not have a potentially damaging Failure Mode and Effect Analysis (FMEA). Written in 1969, it was one of a series of hazard analyses required at the pre-certification stage aimed at detecting potentially dangerous structural or operational weaknesses. Drafted by Convair, a division of General Dynamics, which was building the doors and much of the fuselage, it classified the door latching mechanism as a fourth degree hazard – one which could cause a catastrophic crash.[29] General Dynamics was contractually prohibited from communicating directly with the FAA, so all its memos went directly to McDonnell Douglas. In this case, a vice-president for engineering disagreed with the analysis and conclusions and deleted that section of the FMEA. Even more important, this particular analysis was never sent, even in its altered form, to the FAA[30].

Second, the system of designated representatives outlined above reduced the ability to catch mistakes in assumptions, analyses or conclusions. All the incentives encouraged forward progress, not a time-consuming new design.

The theoretical hazards outlined by the FMEA became a physical reality in the pressure vessel test of fuselage Number Two on 29 May 1970. 'Between 3 and 4 psi the forward cargo compartment door blew out with a violence that tore the door off its hinges and buckled the floor adjacent to the door' (Johnston, 1976: 231). Douglas put the blame on 'human failure', claiming the mechanic had not properly closed the door.[31] Some minor adjustments were made, though none significantly increased the door's safety.

Finally, by 12 June 1972, McDonnell Douglas had received approximately 100 complaints about cargo door malfunctions in airline Maintenance Reliability Reports. Douglas was not required to communicate these to the FAA, and decided it was not warranted.

Around 7.20 that evening American Airlines Flight 96 took off from Detroit Metropolitan Airport on its way to Buffalo and then New York. About 5 minutes into flight, nearing 11 500 feet over Windsor, Canada, an improperly closed rear cargo door blew out, causing rapid depressurisation of the aircraft. Only 50 passengers were on board so there was not an excessive amount of weight on the rear cabin floor,[32] which partially collapsed causing severe damage to the hydraulic lines, but not completely severing them. The flight crew was able to land the plane safely.

Subsequent investigations revealed another serious weakness in the cargo door design: the linkages between moving parts were so weak a baggage handler could 'close' the door (defined by the door handle being in its locked position), even though the locking pins and holes were not properly aligned, by merely exerting a little extra strength. This same action extinguished the cargo door warning light located on the flight engineer's panel so an improperly closed door appeared to be secured. The National Transportation Safety Board recommended the door 'be modified to make it "physically impossible" . . . to be improperly closed', and the cabin floor 'be modified . . . to prevent its collapsing after a sudden decompression' (Eddy *et al.*, 1976: 152). The RLD concurred, and sent a delegation to meet with McDonnell Douglas and FAA officials. In response, McDonnell Douglas noted their doors conformed to all relevant FARs, and the FAA commented, 'We do not concur with RLD views concerning the inadequacy of FAA requirements' (as quoted in Eddy *et al.*, 1976: 162).

The 1969 incriminating FMEA, the 1970 fuselage test blow out, and the 1972 Windsor incident came at very inconvenient moments in the DC10 production and marketing processes. The 1969 FMEA was written just when 'Douglas and General Dynamics were making the rite of passage from designing to building the plane, and Convair was tooling up' (Johnston, 1976: 229). Douglas had begun milling flight deck window frames. Convair had begun making parts, its inventory growing into the thousands. Preparations were underway for the first door mock up.

The DC10 programme pressed on gaining an irresistible momentum. By January 1970, the first fuselage section was on the assembly line at Long Beach and the '270 000 parts were being fastened with two and a half million fasteners' (Johnston, 1976: 231). By April, the wing and fuselage sections had been joined and the multitude of wires, cables, hydraulic and pneumatic lines were being connected.

Everything was on schedule and excitement filled the air as the paper plane came to three dimensional life.

Management concern about timing was embodied in their DC10 production motto – 'Fly Before They [Lockheed] Roll'. Douglas was competing both with Lockheed, to be the first to the market, and with its own delivery deadline.[33] A 'countdown atmosphere' was created with a flood of memos regarding the number of days left to first flight and a new weekly deadline on the aircraft's assembly (Johnston, 1976: 231). By the time of the fuselage test blow out, many of the more than 200 requisite steps for certification had been completed, including some flight testing of fuselage Number One. It was important these steps not be retread; time was of the essence.

These first two inconveniently timed events were dealt with internally by McDonnell Douglas. The FAA did not know about them. The 1972 Windsor incident was public knowledge, however, and, according to procedure, the FAA Western Regional Office began drafting an Airworthiness Directive.

All aircraft require modifications as they enter the stream of commerce. The manufacturer distributes Service Bulletins (SB) to aircraft operators,[34] identifying necessary changes. If these are not related to airworthiness, the operator bears the cost. An Airworthiness Directive (AD) is 'served' on an aircraft only when modifications are safety related and mandatory. Contracts guarantee delivery of airworthy aircraft. Failure to comply with an AD renders an aircraft non-airworthy, so the manufacturer bears the cost.

Service Bulletins and Airworthiness Directives are different in another significant way. SBs are rarely public. ADs are public documents, automatically circulated to news media, aviation attachés of all foreign governments, and FAA overseas offices.

The Windsor incident could hardly have occurred at a more inconvenient time for McDonnell Douglas. June marked the middle of the summer air travel season. If the plane were grounded, there would be serious economic consequences for its airline operators, who would probably demand some form of compensation to cover their losses.

Equally as serious, the summer of 1972 marked the beginning of an intense international sales competition between Lockheed and McDonnell Douglas. By this time most US airlines had made a commitment to one aircraft, now it was important to sell to foreign airlines. McDonnell Douglas and Lockheed sales teams pursued the same strategy. More than actual numbers sold, what counted 'was to obtain first orders from certain airlines thought to occupy influential positions'. A key target was Turkish Airlines because it was ready to purchase new equipment and was 'thought to be capable of influencing other Middle Eastern airlines by its example' (Eddy *et al.*, 1976: 158). Each manufacturer had a specially prepared plane laden with executives, engineers, sales personnel and public relations experts which flew around the world in an effort to sell, sell, sell. McDonnell Douglas's plane, labelled 'Friendship '72', was being readied for departure when the Windsor incident occurred.

The negative publicity associated with an Airworthiness Directive prompted a flurry of calls between McDonnell Douglas President Jackson McGowen and FAA Adminstrator John Schaffer. McGowen convinced Schaffer of corporate concern for the aircraft's integrity. Modification work had already begun. The ensuing 'Gentleman's Agreement' supressed the AD and allowed McDonnell Douglas to determine and incorporate changes which would create a fail-safe cargo door.

McDonnell Douglas had earlier convinced FAA personnel the door was fail-safe and the 'possibility of a door failure was "extremely improbable" 'meaning anything from "one failure in 10 000 flight hours" to one in a billion!' The door blew open within the first 5000 hours of service. 'Failure had not been "*extremely improbable*" ' (Johnston, 1976: 202; emphasis hers). The basis for trusting the McDonnell Douglas judgement or intentions was a bit shaky (see Appendix 5). In the end, measures taken by Douglas engineers did result in a fail-safe door. Unfortunately, all the changes were not completed until months after the 1974 Paris crash, and then only as a result of an Airworthiness Directive issued two weeks after the tragedy. The series of McDonnell Douglas Service Bulletins had not carried enough clout to result in prompt operator compliance.[35]

Even worse, though they had papers saying otherwise, the ill-fated Turkish airliner and a Laker aircraft had not been properly modified. The THY aircraft was on a McDonnell Douglas parking area when the SBs were issued, but had been misrigged. The locking pins travelled less rather than more distance and an additional shim made the flight engineer's warning light go out before the door was sealed. This made the door seem easier to close; all the warnings had been about not applying too much pressure, so for the door to close easily seemed no cause for alarm. Post-crash investigations revealed McDonnell Douglas personnel were to blame for the misrigging.

The Paris Crash

The weather was not perfect on 3 March 1974. The flight was late because re-routings forced by the BEA strike led to the last-minute addition of many passengers and their gear. The chaos brought on by the day's events meant the THY baggage handler was responsible for closing and locking the rear cargo door, rather than either the mechanic or the baggage handler's supervisor who usually checked the plane. The peephole, added after the near disaster over Windsor, was rendered useless as a means of determining proper closure because of the rush, the weather, and the fact that the explanatory decals, also added after Windsor, were written in English. The baggage handler was well trained and fluent in three languages, unfortunately not English. Also, since this aircraft had been misrigged, the improperly sealed door turned off the flight engineer's warning light. So, as the crew went through the check list, cargo door closure was acknowledged, and the aircraft taxied away from the gate with one crucial function not adequately performed. When the plane reached 11 500 feet, the pressure blew out the rear cargo door. The resulting sudden decompression collapsed the cabin floor and sucked out two rows of seats with their six passengers. This severed all three hydraulics cables, so the flight deck crew could no longer manipulate the aircraft's control surfaces. The autopilot was still available since its cables were secured in the cabin ceiling, but its effectiveness lay in its ability to activate and 'steer' the hydraulics system – and there were no hydraulics to use.

The National Transportation Safety Board identified several probable causes of the Paris crash, including poor cargo door design, redundant hydraulics cable placement under the cabin floor, failure to look in the peephole, incomplete modifications along with misrigging the linkages and adding an 'alien shim', and failure to follow-up with corrective measures after Windsor. The FAA was implicitly culpable because at least three causes were its responsibility – cargo door design, cable placement and post-Windsor changes.

Legally, however, the government cannot be held liable for negligence in its performance of its aircraft certification function (see Appendix 6). The Supreme Court has consistently ruled the FAA 'cannot be sued for carelessly failing to find a defect that causes a commercial air disaster . . . The FAA has a statutory duty to promote safety, not insure it' (Epstein, 1984: 15A). The Court has been careful not to 'arrogate to themselves the [FAA's] adjudicative function' and has accepted the argument that 'it would be physically impossible to establish regulatory standards' capable of preventing any accident. And, even if it were, the cost 'would economically paralyse the aviation industry. *Such over-regulation would be contrary to the Administrator's responsibility to foster the development of civil aeronautics* (Harrison and Kolczynski, 1978: 43; emphasis mine).

Two days after the Paris catastrophe, the new FAA Administrator Alexander Butterfield created an *ad hoc* committee to examine the DC10's evolution and the FAA's role in it. A 6-week investigation produced a 'damning indictment of the DC10 door design and of the regulatory system in the United States that had permitted its certification' (Eddy *et al.*, 1976: 160).[36] The Subcommittee on Investigations of the House of Representatives' Interstate and Foreign Commerce Committee conducted its own review and concluded that 'between June 1972 and the Paris crash of March 1974 "through regulatory nonfeasance, thousands of lives were unjustifiably put at risk"' (Eddy *et al.*, 1976: 160–1).

An addendum to this discussion is necessary regarding the role of domestic politics in this whole affair: during the first six months of 1972, McDonnell Douglas and its executives contributed over $90 000 to Richard Nixon's re-election campaign. All the money was publicly declared and most of it paid before Windsor, thus it cannot be said that any clandestine 'deals' were made to buy favourable treatment for this particular incident. None the less, there is room for speculation on the basis of a Charles Lindblom (1977) argument regarding the privileged position of business in American politics.

The O'Hare Crash

Ironically, the 1979 catastrophe involving a US-manufactured aircraft, operated by a US airline, loaded with US citizens, departing from a US airport with a US destination and crashing on US territory, was ultimately to have greater global ramifications than the 1974 tragedy involving a US manufactured aircraft, operated by a Turkish airline, loaded with citizens from 21 different countries, departing from a French airport with a London destination and crashing on French soil.

The 1979 tragedy spawned several extensive reviews of Federal Aviation Administration regulation creation, certification and maintenance and operations oversight procedures. Their conclusions and recommendations focused on the inextricable interconnection of these processes and revealed that many of the weaknesses which had contributed to the 1974 disaster were still affecting the international commercial aviation safety regime. Between 1974 and 1979, the international aviation community made no attempt to alter the FAA's predominant role in the regime. In 1979, however, the European aviation community sought its own resolutions and recertified grounded aircraft several days before the FAA.

Accident Investigations

According to standard procedure, the National Transportation Safety Board and the Federal Aviation Administration investigated the cause of the 25 May 1979 disaster. Reflecting the fundamental problems associated with achieving aviation safety, the process was hampered by inadequate resources, poor communications and data collection procedures, and powerful political-economic incentives to avoid 'over-regulation'.

Three days after the crash a quarter-inch bolt used to attach the engine and pylon mounting to the wing was discovered on the taxiway travelled by the crashed aircraft. Thinking a fractured bolt might have caused the disaster, the FAA ordered US airlines to ground their DC10s for inspections. Foreign airlines operating 140 DC10s were admonished to follow suit. Within 3 hours the planes were released only to be grounded again the next day so the entire engine mounting assembly could be inspected. Spot checks had revealed 'grave and potentially dangerous deficiencies' in several pylon mounting assemblies ('FAA Orders US . . . ', 1979: 2–3). The problem was more complicated than just a failed mounting bolt.

By 1 June the NTSB's chief metallurgist determined the bolt broke from the sudden stress associated with the crash, rather than from wear-induced fatigue as first thought. Twenty-four hours later, five planes were selected for more thorough inspections. Then on 6 June the FAA issued an emergency order suspending the DC10 airworthiness certificate because cracks, similar to those believed responsible for the structural failure, were discovered in two aircraft which had been cleared for flight. The order applied to all DC10 landings and takeoffs within the US and effectively grounded all 278 DC10s in service throughout the world.[37]

As the FAA and the NTSB searched for the crash's cause and acceptable solutions, grounding costs began to mount. Daily lost revenues were estimated to be $15–20 million, with the burden falling discriminately on airlines with DC10s in their fleets. For example, Continental DC10s carried 41.5 per cent of its passengers and it lost about $900 000 in daily revenues during the grounding. Western flew 26 per cent of its passengers on DC10s; it lost about $600 000 daily. In addition, each suffered losses in the stock market: Continental's earnings fell from $3.03 in 1978 to less than $1.00, while Western's fell from $2.18 to $1.50.

Labour expenses were also costly. DC10s were flown by senior pilots contractually protected from being furloughed;[38] so airlines were paying employees for no work. Some, like Western, were able to stop paying salaries under such conditions, but all other benefits remained in place.

Indirect costs rose, too, as travellers making summer plans booked on airlines which did not fly DC10s (Carley, 1979: 1).

The National Transportation Safety Board *Accident Report*, released 21 December 1979, identified the accident's probable cause as maintenance-induced damage leading to the 'separation of the No.1 engine and pylon assembly at a critical point during takeoff'. This caused 'the asymmetrical stall and the ensuing roll of the aircraft' out of which the flight deck crew could not fly. 'Contributing to the cause of the accident were the vulnerability of the design . . . to maintenance damage' and deficiencies in FAA surveillance, reporting and communications systems (National Transportation Safety Board, 1979: 69).

Once again, it began with the regulation creation process. The NTSB concluded that the certification of the DC10 was carried out in accordance with the rules in effect at the time, which were based on prevailing engineering and aeronautical knowledge and standards. However, the regulations were inadequate because 'they did not require the manufacturer to account for multiple malfunctions resulting from a single failure'. McDonnell Douglas considered structural failure of the pylon and engine to be 'unacceptable', so they were 'designed to meet and exceed all the foreseeable loads for the life of the aircraft'. Thought to be as vital to flight as a wing, McDonnell Douglas did not perform an analysis based on their loss. Logic supports the decision not to analyse the loss of a wing, because 'further flight is aerodynamically impossible'.

> However, similar logic fails to support the decision not to analyse the structural failure and loss of the engine and pylon, since the aircraft would be aerodynamically capable of continued flight. The possibility of pylon failure, while remote, was not impossible. Pylons had failed. Therefore, fault analyses should have been conducted to consider the possible trajectories of the failed pylon, the possibilities of damage to aircraft structure, and the effects on the pilot's ability to maintain controlled flight (National Transportation Safety Board, 1979: 58).

And, again, critical failures occurred during the design phase of certification, when inadequate resources required FAA dependence on designated representatives. Too few analyses and studies were reviewed, and these did not delve deeply enough into the underlying assumptions supporting conclusions. In addition to the design problems discussed above, four more contributed to the 1979 crash.[39]

First, McDonnell Douglas, following in the tradition of its earlier jet transports, placed wing control hydraulics lines along the leading (front) edge of the wing. 'The only reason for having hydraulics lines on the leading edge is that it is simpler and cheaper to put them there' (Newhouse, 1982: 101).

Second, DC10 slat control mechanisms were not fail-safe. Slats are movable wing sections controlled by hydraulics. These extend to provide extra lift for takeoff and landing, the two most critical periods of flight. Sudden slat retraction can destroy aircraft lift capabilities, causing it to enter into a stall and precluding pilot control. The FAA considered the question of uncontrolled slat asymmetry during certification, but the manufacturer only had to demonstrate that the aircraft could stay aloft while already airborne. No tests were conducted at the critical points of rotation or touch down (Green *et al.*, 1979: 1; United States House of Representatives, 1980c: 22–6).

Third, the DC10 did not have redundant stall warning and slat disagreement, often a prelude to stall, devices.[40]

Fourth, because engines are not supposed to fall off aircraft, there was no flight deck instrumentation to indicate such an occurrence.

Tragically this was not a 'forgiving design tolerant of failure', rather, it was vulnerable to maintenance damage. Failures in FAA maintenance and oversight duties contributed substantially to the 1979 disaster. This time another layer of FAA dependence on designated representatives – airline Principal Maintenance Inspectors (PMIs) – played a significant role.

Ironically, the series of events leading to the crash began out of a concern for safety. In March 1975, five airlines reported cracks in a major pylon bearing, a 'monoball', which McDonnell Douglas believed could lead to structural damage of the wing pylons. In October it issued a Service Bulletin recommending inspection and replacement if cracks were found.[41] Replacement required engine and pylon removal. The Service Bulletin repeated instructions for this time-consuming process already contained in three versions of the DC10 service manual: the 8000 pound engine and 4000 pound pylon were to be removed separately. In 1977, an American Airlines supervisor had to complete maintenance on a Varig Brazilian Airline DC10 under contract and could see he was running out of time. He thought of removing the engine and pylon as a single unit using a fork lift. This would halve the time required (Carley, 1979: 25; Green *et al.*, 1979: 1; Rempel, 1979: 1).[42] The technique was observed by McDonnell Douglas personnel twice[43] and no objections were made. Service manuals still recommended, and McDonnell Douglas and most other airline mechanics still followed, the separate removal process. Neither the FAA nor the NTSB knew of the procedural change until after the 25 May 1979 crash at O'Hare.

The 1978 introduction of US airline economic deregulation had the immediate effect of pinching historically slim profit margins (Golich, 1986, 1988). The imperative to cut costs took an even greater meaning. Aircraft do not generate revenues while on the ground, so discretionary maintenance is often eliminated, and any shortcuts in mandatory maintenance are diligently pursued. According to testimony before the House Government Activities and Transportation Subcommittee by representatives from the Aircraft Mechanics Fraternal Association and the International Association of Machinists and Aerospace Workers, air carriers consistently 'and dangerously deferred needed maintenance and emasculated their quality control departments' (United States House of Representatives, 1980c: 459).

Unfortunately, the American Airlines engineers who designed the new maintenance procedure did not adequately evaluate 'the capability of the fork-lift to provide the required precision for the task, or the degree of difficulty involved in placing the lift properly, or the consequences of placing the lift improperly' (National Transportation Safety Board, 1979: 68; see also United States House of Representatives, 1980c: 52). Even worse, maintenance personnel deviated from their own prescribed procedure by removing the pylon bolts in reverse order, which increased pylon vulnerability to damage (National Transportation Safety Board, 1979: 68; see also Carley, 1979: 25). It was virtually impossible for the FAA to discover maintenance deviances because ' "the lion's share of maintenance necessarily takes place in the midnight shifts, long after the 8-to-5 working (federal) inspectors have gone home" ' (former ALPA President John J. O'Donnell as quoted in Green *et al.*, 1979: 1).

Unfortunately, the regulators were unable to make an informed decision regarding pylon problems because neither the manufacturer nor the operators had reported the difficulties they were experiencing. On two separate occasions,[44] Continental Airlines personnel noted the engine-pylon mounting assembly was damaged during maintenance. The damage was similar to that inflicted on the ill-fated American Airlines aircraft. The carrier classified the cause as maintenance induced and notified McDonnell Douglas. Neither carrier nor manufacturer interpreted FARs to require further investigation or notification to the FAA. McDonnell Douglas did notify other DC10 operators, but only via a low priority Operational Occurrences Report in a paragraph below the description of a flight attendant's toe injury resulting from a food cart. No specific warning was issued regarding potential hazards of this method of pylon and engine removal. Although the bulletin was signed by a McDonnell Douglas DER, the FAA was excluded from this flow of information.

In addition, the information flowing to the FAA was not organised in a fashion which would allow warnings to be the reasonable interpretation. Between 1971 and May 1979, at least 53 pylon modifications were recommended in Service Bulletins, and 68 instances of pylon-related trouble were reported to the FAA by US airlines. Routinely entered into computers at the FAA records centre, someone decided, ' "This one isn't too important" or "This is not a trend" or "This is not crucial" ' (Basnight as quoted in Dean and Kendall, 1979: 3). Although the rate of problems with DC10 pylons was nearly double that with B747 pylons and greatly exceeded the L1011's one reported problem, FAA personnel did not feel there was sufficient data 'to indicate a worrisome pattern' (Green *et al.*, 1979: 1).

The reasons support Olson's logic of collective action and reveal a complacency regarding aviation management. It was not in the perceived interest of industry participants to inform the FAA: there was neither a positive incentive nor a negative sanction which would have impelled them to do so. Instead, there was a perceived negative sanction associated with informing the FAA: it might have led to more stringent surveillance or increased maintenance. Not only is this a nuisance, it can be very costly. The FAA, suffering from its dual responsibility role and lack of resources, was lulled by the argument that safe operations were in the industry's own best interest and no stronger incentive could prevail.

FAA behaviour subsequent to the O'Hare crash provides further evidence of the poor quality of its oversight duties and intensifies the irony surrounding the whole affair. The 2-inch cracks which finally led to the 6 June grounding were considered serious because they had not been previously detected, and the aircraft which had them had not undergone the kind of maintenance procedures then thought (and later proven) to have caused this type of damage. Actually the cracks had been there all along, but the FAA's first mandated action was to require only visual inspections of the critical pylon area. Only several days later did the FAA also require thorough 'non-visual' (x-ray) inspection of these sane components, and it was then the cracks were discovered. Apparently, these actions were motivated by a desire to avoid imposing an excessively expensive series of inspections on the industry. However, had the x-ray procedure been initially prescribed, the 'new' cracks might not have led to the very costly grounding which followed.[45]

As with the 1974 catastrophe, the FAA did require some changes subsequent to the 1979 crash. Airworthiness Directives were issued in early 1980 which required increased redundancy in the DC10 stall warning system, strengthened pylon attachments to prevent separation from the aircraft, and a new maintenance procedure to prevent weakening pylon attachments.

A 6-month study formed the basis for the proposed changes, but otherwise concluded 'the DC10 pylon is "fundamentally sound" and can serve 73 000 flight hours without failure unless it is damaged during maintenance or by some other cause' (as reported in 'FAA Proposes . . . ', 1980: 26). Written by McDonnell Douglas, the FAA and Air Force reviewed the study, but no full-time government supervisors were assigned to the firm during its work. The chance to review assumptions and methodologies and their bearing on conclusions was once again eliminated. An agency official defended the decision: 'Douglas knows the airplane. The FAA doesn't have enough people to get it done in a reasonable length of time' ('FAA Proposes . . . ', 1980: 26).

European Response

European aviation authorities and airline executives, dissatisfied with the investigation's slow progress, blamed politics: Umberto Nordio, Chairman of Alitalia, suggested the

> American government had taken the action for political reasons, to appear as a defender of consumer interests. The grounding order was hurried and superficial . . . The US government should have taken less severe temporary measures to check the safety of the planes. (Carley, 1979b: 29)

Similarly, Reinhardt Abraham, a member of Lufthansa's management board, commented that 'US authorities may make "political" decisions on the safety of the DC10 rather than decisions based on the technical merits' (Carley, 1979b: 29).

These officials were concerned about the economic health of European airlines heavily dependent on the DC10. In Europe, airlines like the UK's British Caledonian and Laker, Switzerland's Swissair and France's UTA operated DC10s almost exclusively. Thirteen European airlines, operating 58 DC10s lost an estimated $40 million during the grounding ('Europeans Clear Inspection Program . . . ', 1979: 25).

On 2 June 1979 delegates from the Association of European Airlines (AEA) and the European Civil Aviation Conference (ECAC) met in Strasbourg.[46] They agreed to design a programme of inspections and maintenance to ensure and preserve DC10 airworthiness so it could be returned to service. A special task force comprised of officials from Alitalia (representing ATLAS), British Caledonian, and Swissair (representing KSSU) met under the aegis of AEA's Technical Affairs Committee. Three days later they submitted their programme to a meeting attended by AEA and ECAC aircraft maintenance and engineering experts. All DC10 operators, their airworthiness authorities, McDonnell Douglas and the US FAA had been invited. The Europeans hoped the FAA would give tacit blessing to their programme, but it refused, claiming it had no legal authority to approve or disapprove. The programme was formally reviewed and approved by an Extraordinary Maintenance Review Board on 18 June and European DC10s were flying again the next day. The FAA gave permission for these aircraft to fly over US air space but takeoffs and landings were still prohibited.

In an effort to press their case, ECAC officials met with US representatives in Paris on 25 June 'to discuss the urgent wish of ECAC States to resume DC10 operations to the United States' (European Civil Aviation Conference, 1979d). Delegates reviewed the status of the US investigation and the conclusions reached by European authorities. They considered the legal aspects of resuming DC10 operations with respect to both national laws and international agreements. At a 4 July meeting in Washington DC,[47] ECAC argued the United States was obligated 'both under the Chicago Convention on international civil aviation and under their bilateral agreements with ECAC Member States, to authorise European carriers to resume DC10 operations to the United States'. But their formal request 'to allow such operations' was rejected (European Civil Aviation Conference, 1979c).

Largely on the basis of Europe's inspection and maintenance programme, several international airlines resumed DC10 flights.[48] Japan was virtually the only country relying on FAA clearance for resumption of DC10 service.

Europeans took action for five reasons: 1) The cost of keeping aircraft grounded was high. 2) European carriers operated a different model DC10 than those flown in the US.[49] In all the mandated inspections no pylon cracks similar to those found on US aircraft were discovered on European models. 3) European airlines had conformed meticulously to McDonnell Douglas and FAA prescribed maintenance procedures, whereas American airlines had not. 4) The initial mishandling of the accident investigation by the FAA and NTSB led to serious doubts among European authorities regarding procedural professionalism. 5) These doubts were intensified by revelations that FAA culpability in 1979 was related to many of the same weaknesses which had contributed to the 1974 crash.

Europe's response to the 1979 disaster was remarkably cohesive. Airworthiness authority members agree that previous collaborative work toward developing Joint Airworthiness Requirements (JARs) was helpful when the time came to design a DC10 recertification programme.[50] Between 1974 and 1979, the JAR consortium made steady progress establishing type certification standards and convincing European Community members to adopt them as their own (Ashford, 1986). Meanwhile, Airbus aircraft were successfully penetrating many former US export markets, perhaps signalling a change in international air transport dominance.

CONCLUSION

The 1974 and 1979 DC10 crashes revealed deadly weaknesses within the international civil aviation safety regime. In 1974, the long, involved and costly process of regulation creation impelled industry and government to use obsolete regulations to judge the safety of a new technology aircraft. In 1979, the long, involved and potentially costly results of submitting a new maintenance procedure for thorough scrutiny discouraged the kind of communications among relevant participants which might have prevented the O'Hare tragedy.

Fatal vulnerabilities existed in regulation creation, certification and maintenance and operations oversight procedures within the nearly omnipotent United States Federal Aviation Administration. Because so much of the world relied on the FAA to ensure commercial aviation safety, these weaknesses affected people and states far beyond US national borders. The areas of vulnerability were particularly susceptible to political and economic pressures.

Tragically, many of the problems which led to the Paris accident, contributed to the O'Hare tragedy. Little progress had been made during the intervening five years toward rectifying the fundamental vulnerabilities of the safety regime. In spite of the horrific nature of the 1974 disaster, little had been done outside the US to decrease dependence on US hegemony. In part, this can be explained by the incredible costs related to the development of a commercial class aircraft industry. If the solutions were seen to be the creation of an independent set of standards, procedures and personnel to evaluate aircraft airworthiness issues, this would add substantial costs to an already very expensive product. Most states could not afford either solution, and so, as in 1974, in 1979 the FAA was essentially responsible for guaranteeing the safety of the global air transport network.

The reasons why fundamental weaknesses in FAA procedures persisted were related to domestic and international political–economic factors. The need to promote the industry prevailed over the need to regulate it because of its strategic value. Budget deficiencies and lack of public support for increased expenditures prevented the quantitative or qualitative upgrading of the FAA staff or its information system. Complacency about an adequate aviation safety regime combined with bureaucratic inertia and procedural rigidity to create an attitude of arrogance impervious to new knowledge regarding the effects of technological innovations or structural changes in the global and domestic decision environments.

By 1979, however, European state and corporate elites had learned that collaboration was the key to autonomy from US hegemony. They flexed their stronger political–economic muscles and challenged existing regime rules and decision-making procedures by calling for new bilateral air transport agreements and by recertifying and flying their DC10s before the FAA did. They embraced new production technologies and transnational management skills to reinvigorate their commercial class aircraft manufacturing sector. (This would challenge regime principles and norms). Finally, they transferred their knowledge about gains reaped from co–operation in the private sector to the political arena by continuing to strengthen their transnational Joint Airworthiness Requirements consortium. Learning that benefits of co–operation can outweigh potential costs may be a key lesson for the future of global relations.

5 Conclusions: Structure, Process and Reciprocity

> Knowledge creates a basis for co-operation by illuminating complex interconnections that were previously not understood . . . (Ernst Haas, 1982: 23)

> In the international arena, neither the processes whereby knowledge becomes more extensive nor the means whereby reflection on knowledge deepens are passive or automatic. They are intensely political. (Friedrich Kratochwil and John Gerard Ruggie, 1986: 773)

Disasters and human tragedy capture our attention in a compelling way. The unfolding tale of the 1974 and 1979 DC10 catastrophes dramatically reveals the connections between system and unit levels of analysis, between micro and macro actors and variables, and begin to tell us something about the process of change in our world. These catastrophes were the result of linkages between micro actors – corporate engineers and government bureaucrats working on narrowly defined projects – and the larger forces of change in international power and market structures. The prevailing international civil aviation regime, as an intervening variable between international and domestic structures and processes, both influenced and was influenced by the nature of these linkages.

Virtually all recent regime theory literature speaks of the need 'to develop interactive models that link domestic and international politics more closely' since ' "domestic" policy issues spill over into international politics and "foreign policy" has domestic roots and consequences' (Haggard and Simmons, 1987: 515–16).[1] The lack of detail in system level theories has led to poor explanatory or forecasting abilities (see Nye, 1988). By adding domestic dynamics scholars run the risk of losing parsimony, but potentially gain a more accurate reflection of reality (see Gowa, 1986, 1985; Ikenberry, 1986; McKeown, 1983). Trying to understand the reciprocal dynamics taking place in and between both system and unit levels of analysis presents an even greater challenge.

Scholars have identified at both the system and unit levels key variables which influence actor behaviour. We need to understand how these influence state elites operating at the 'intersection' of domestic and international politics as they select policies designed to serve their conceptions of national interest (Ikenberry, 1986). The story told in this book reveals the inextricable linkages among actors and variables. By doing so it adds to our comprehension of 'process', the ways in which elements of international relations relate to each other (Nye, 1988: 249).

At the system level power and market structures are critical. Equally as important are non-structural variables such as changes in reigning ideas and knowledge, technological innovation, and shifts in the number and patterns of transnational interactions (Nye, 1988: 250; Odell, 1982).[2]

At the domestic level, policy makers are influenced most fundamentally by ideology, which shapes public perceptions about appropriate state-economy relationships and about which and how issue areas should be managed by governments. Wielding special influence, bureaucrats determine the 'short list' of feasible policies; they bias policy effects through interpretation and implementation (see Chapter 4). Private sector demands and actions are becoming increasingly important.[3] Sometimes influence is exercised consciously and formally by economic interest group associations lobbying for special treatment. Policy can also be influenced less consciously, but no less powerfully, when corporations pursue survival strategies which place constraints on and create opportunities for state behaviour.[4] Finally, policy selection will be a function of the limitations and possibilities inherent in prevailing regimes.

Many scholars have noted that regimes, like other institutions, resist change.[5] Organisational inertia and the effects of prior policy choices are partially responsible for this phenomenon because state and non-state actors have learned how to operate both internationally and domestically within familiar parameters. Additionally, the positive contribution of regimes to reducing uncertainty and to facilitating global management of a wide-ranging number and type of issue areas means regimes persist without supervision by a central authority (Crawford and Lenway, 1985: 37).

Nevertheless change does occur. State and non-state actors adjust to changes in domestic and international environments. They alter their beliefs and actions incrementally as a result of daily interactions or on the basis of new information. They 'develop knowledge or skill by study or experience', such as feedback about which policies worked and which failed (Keohane and Nye, 1987: 749). Knowledge, ' . . . the sum of technical information and of theories about that information which commands sufficient consensus . . . to serve as a guide to public policy . . . ' (Haas, 1980b: 367–8), is affected by 'changes in reigning ideas' (Odell,1982: 58).

Communication provides opportunities for state and corporate elites to learn, to redefine interests, and to select new policies (Nye, 1988: 250). The exchange of ideas, knowledge and technological innovation can lead to regime creation and, later, to new ways of behaving within regime parameters. New rules and decision-making procedures may be adopted. Eventually, critical variables may be so altered that new or modified regimes emerge guided by different norms and principles. Crises are often the necessary catalyst for learning because they force decision maker recognition of structural and non-structural changes. The process of change never ceases.[6]

The thesis of this book is that macro-level changes are affected by the daily behaviour of an aggregate of micro actors operating according to domestic and international regime structures which delimit appropriate political and economic action at both levels. A simple process model looks like this:

Domestic & international structural & non-structural variables → **influence micro actor behaviour**

Micro actor behaviour influences decision-maker perceptions of what national interest is and how to serve it best. The foundations for these perceptions change gradually. A model of policy selection and implementation is more complex:

Policy selection & implementation = **a function of regime parameters + domestic ideology + changes wrought by micro actor behaviour**

Decision makers must be aware of and receptive to changes in ideas and knowledge and to environmental changes in order to adjust policies. Awareness is generally triggered by a crisis because previous policy choices have generated complacency regarding their appropriateness. Receptivity is a function of ideology and the pressures created by micro-level activity and demands. Adjusting policy selection and implementation adds a dimension of learning:

Adjusting policy selection = & implementation	**learning, a function of receptivity to and awareness of changes in reigning ideas and knowledge, technological innovation, and the number and patterns of international interactions**

We cannot presume that all learning results in the selection and implementation of policies leading to Pareto optimality. Government and corporate elites can be misinformed, select myopic, destructively conflictual policies, or implement policies poorly as well as learning, selecting and implementing 'good' policies 'well'. Because of the shifting and often conflicting demands to which decision makers must respond, policy selection and implementation can be influenced by 'inappropriate' pressures. Events surrounding the DC10 production, certification and crashes support this contention. Concerns about declining political, military and economic power affected decisions about technologically solvable safety problems. Fortunately actors can learn from failures as well as successes.[7]

REGIME CREATION

Regime creation involves the establishment of incentives which impel or compel, however imperfectly, actor compliance. In its original version the hegemonic stability paradigm suggested that regimes were created by altruistic hegemons to manage public goods-type issue areas such as security, trade and monetary relations. The hegemon was willing to bear extra costs to gain more broadly conceived and long-term benefits associated with stability (see for example, Kindleberger, 1973; Keohane and Nye, 1977).

Recently several scholars have cautioned that we need to 'think about incentives facing the potential hegemon . . . ' (Keohane, 1984: 39). Hegemons may not be altruistic, rather they may reap immediate and tangible benefits for leadership *and* pass on costs to other regime participants (see Snidal, 1985: 587). Equally important, 'theories of hegemony should . . . explore why secondary states defer to the leadership of the hegemon' (Keohane, 1984: 39). Initially reluctant, secondary states 'receive net benefits', so they are willing to 'recognise hegemonic leadership as legitimate and to reinforce its performance and position' (Snidal, 1985: 587). Regime 'principles and norms inculcate an anticipation of reciprocity' which allows secondary states to accept subordinate level participation as being in their national interest (Smith, 1987: 256).

The evidence herein suggests that participants co-operate because they recognise that 'the costs of national self-reliance are excessive' (Haas, 1980b: 35). Although absolute costs and benefits may be asymmetrically distributed, regimes are designed to 'incorporate some of the aims of the dominated (Hart, 1986: 14) since 'actors are more likely to agree on a regime, the more they win and the less they risk by co-operating, on the one hand, and the less they win and the more they risk by defecting, on the other' (Jönsson, 1987b: 2).[8]

The effects of prior policy choices also influence state elites as they design regimes. First, elites from more powerful states will seek to retain those elements which helped their state achieve an influential position. Second, state elites will seek to avoid destructive policies. In the case of most post-Second World War regimes, international civil aviation included, state elites were determined to avoid policies which had led to global conflict (Gaddis, 1972). They had learned that some level of 'international collaboration [was] *required* for the realisation of national goals' (Haas, 1980b: 374; his emphasis).

Chapter 2 confirms these assertions. In 1944, a hegemonic United States created an international civil aviation regime which included safety regulation and benefited everyone. Safe, reliable, global air transport was necessary for the development of trade and communications networks perceived to be vital to state and corporate survival. The guiding principle of the regime remained air space sovereignty. The United States stood to gain the most. It would dominate civil aviation by virtue of market size, technological prowess and skills and assets related to wartime long-range transport responsibilities. US state elites wanted to facilitate aviation's development because they believed it would benefit the domestic economy, thereby strengthening an important power resource base. The United States bore the initial costs of creating and enforcing regulations but these had been demanded by American citizens anyway. The standards creation process was co-ordinated with aircraft production schedules, decreasing costs and facilitating reasonably priced products. Fungible resources could be used to build better aircraft. Since no airline in the Western world was profitable without access to the US market, it was fairly easy to transfer costs to other states and their corporations by requiring them to meet US standards.

Although secondary states were initially disappointed because their preferences for creating international air transport as a public utility and having universal standards set by an international governmental body had not prevailed, there were enough positive incentives for them to acquiesce. States retained several familiar benefits: they could control access to their territory, generate revenues, employ people, stay at the leading edge of technology and increase prestige. As Stephen Krasner (1985) points out, in an international system marked by extreme asymmetry in power capabilities among states, a regime based on 'authoritative resource allocation' gave secondary states more control over their destinies than a 'market oriented' regime would have done.

Micro actors also benefited. Consumers enjoyed safer, more reliable air transport than that which would probably have been provided by corporations alone. They gained lower fares because decreased costs accompanied the efficiencies intrinsic to standardisation, the first level of safety efforts (see Chapter 2; Golich, 1988). Airlines gained because safer service attracted more consumers, until eventually they were won over to air travel for distances over 200 miles. Economic regulation meant airlines could pass on extra costs – safety-related or otherwise – to the consuming public and maintain a secure, if small, profit margin. Finally, aircraft manufacturers gained because the state paid the cost of technological innovation and helped eliminate competition via the use of standards.[9]

Adoption of a regime which embraced both state sovereignty and a regulatory enterprise treatment of an industry servicing an international market constituted United States cultural hegemony by accepting a liberal notion of separation between (and separability of) politics and economics (Russett, 1985: 229). Although not immediately apparent, the stage was set for an increasing tension between a unified global market for civil aviation goods and services and an anarchic state political system. In spite of the liberal framework, 'the political-economic security of the state' was not always perceived to 'be best served' by the 'free' operation of the market. Political choices 'involve a difficult trade-off among different values and value-laden objectives . . . '; they are not always rational or made in the best general (collective) interest (Strange, 1985: 239). Finally, 'decision makers experience incentives to incur long-term costs (payable after election) to show short-term gains, or at least to avoid short-term losses' (Russett, 1985: 227). Corporate elites experience similar incentives.[10]

The regime worked because the United States used 'reward and coercion to maintain' it (Cowhey and Long, 1983: 160) and because its parameters reflected international power and market structures and were shaped by principles and norms acceptable to relevant actors. Regime participants then used their power resources to make the operation of the regime benefit them as much as possible.

REGIME CHANGE

The issue of regime change is particularly intriguing. Created and maintained by hegemons to serve their egoistic self-interest, while incorporating the interests of weaker states, why should regimes change? The hegemonic stability paradigm suggests it is a result of shifts in underlying power structures (Keohane, 1980: 132).[11] This begs the question of what causes the shifts. Herein it is argued that the daily behaviour of state and non-state actors, guided by regime dictates, creates new constraints and opportunities for international relations which eventually lead to changes in both internal and external environments. The process of constant, incremental adjustment eventually generates such tension between 'old' principles, norms, rules and decision-making procedures and the newly prevailing structural and non-structural limitations and incentives that change is imminent. The incompatibility may not trigger change immediately. Often some identifiable crisis is the necessary catalyst. Once elites recognise that adjustments are needed, regimes may facilitate the communication leading to their modification, because they 'link states together' and shape reciprocal behaviour which 'can affect the conceptions of self-interest' (Keohane, 1986: 25). The latter influence the selection and implementation of adjustment policies.

The evidence presented here also suggests that structural shifts can be the result as well as the cause of state and corporate elites learning new ways to act. After the Second World War regimes were designed to redistribute benefits away from the hegemon to other participants. In addition, some 'old' resources became obsolete; some increased in intrinsic value; some increased in value as elites learned to bargain more effectively with them. Also new resources emerged.[12] Structures shifted as a result of policies selected and implemented to fit the parameters of these new regimes. Over time, the market structure changed from a series of national markets connected through trade to one more universally perceived as integrated and global. Micro actors adjusted their behaviour to fit these political and economic shifts, which eventually created new constraints, opportunities and perceptions.

Chapter 3 supports Susan Strange's argument that the crisis of surplus capacity, present when 'demand is insufficient to absorb production at prices high enough both to maintain employment and to maintain profitability for all the enterprises engaged' (1979: 304,fn.2), triggered awareness by state or corporate elites that policies needed to be adjusted. It also supports the arguments that 'the combination of falling hegemony and surplus capacity' produces regime change, and that, within an international regime of freer trade, the initial move may be away from liberal and towards more protectionist behaviour (Cowhey and Long, 1983: 186–7).[13]

Complex interdependence, 'situations characterised by reciprocal effects among actors in different countries' (Keohane and Nye, 1977: 8) intensifies the effects of political and economic structural changes.

> Interdependence generates classic problems of political strategy, since it implies that the actions of states, and significant non-state actors, will impose costs on other members of the system. These affected actors will respond politically, if they have the capacity, in an attempt to avoid having the burdens of adjustment forced upon them. From the foreign policy standpoint, the problem facing individual governments is how to benefit from international exchange while maintaining as much autonomy as possible. From the perspective of the international system, the problem is how to generate and maintain a mutually beneficial pattern of cooperation in the face of competing efforts by governments (and nongovernmental actors) to manipulate the system for their own benefit. (Keohane and Nye, 1987: 730)

Once the hegemon's elites perceive that they have lost the ability to pursue state objectives unilaterally, they may attempt to reassert dominance. Other participants may seek to preserve the regime which redistributed significant benefits to them, or to restructure the regime to increase or speed up benefits accrual. Each will make trade-offs to try to maintain a significant level of participation within a critical international regime. As they adjust, decision makers may learn that co-operation generates benefits unavailable through conflict. Haas notes 'collaboration becomes conflictual only when the parties begin to disagree on the distribution of benefits to be derived' (1980b: 362). The perceived high cost of conflict may also impel collaboration.

Changes in the international civil aviation regime affirm the foregoing assertions. By the 1970s, the internal and external environments within which international commercial aviation operated had changed dramatically. Power resources were more equally distributed. Virtually all state elites sought participation in civil aviation, considered vital to the pursuit of many political and economic policy goals (see Chapter 3). Surplus capacity in both the service and goods segments of commercial air transport created tremendous pressures among private sector domestic groups, namely corporate elites, for some kind of government intervention to preserve the affected industry. This was aggravated by the fact that the industry sectors demanding government assistance were 'more numerous and negotiations more complex' (Strange, 1979: 308). Insecure political elites sought to balance the need to guarantee 'security from external attack' against the need to maintain internal order by contending with interest groups, each 'claiming a measure of justice as the price of their continued support' (Strange, 1985: 236–7). Regime linkages added to the frustration.

Government decision makers had to respond to other domestic demands as well, while trying to regain political and economic dominance in the international system (see Keohane, 1980; Cowhey and Long, 1983: 161; *inter alia*). Americans associated commercial aviation with the military industrial complex. Weary of the Vietnam conflict and rampant inflation back home, they demanded more direct benefits for social welfare programmes and conservation efforts and less support for civil aviation.[14] Unable to 'prime the pump' of commercial aircraft manufacturers with significant financial support for fear the people would not return them to office, state elites chose to support the sale of the All-American DC10 over the L1011 which used British-built Rolls Royce engines.[15] The unofficial 'Buy American' policy followed the tradition, banned by the General Agreement on Tariffs and Trade, of official policies restricting fleet purchases to domestically produced aircraft.

American consumers had grown comfortable and confident in

the level of safety achieved within the industry. Complacency and lack of resources led to relaxed interpretation and implementation of safety regulation, which could be considered subtle, if ineffective, protectionism. The 1972 Gentleman's Agreement between the Federal Aviation Administration and McDonnell Douglas was a dramatic example. Lack of resources meant government scrutiny of the aircraft manufacturer was not as thorough as it might otherwise have been. This led to the nearly disastrous Windsor incident and other physical manifestations of theoretical warnings. Complacency, characterised by satisfaction with private sector concern with safety regulations, meant government officials trusted that the aircraft manufacturer would correct the mistake. The result was the 1974 catastrophe.

Chapters 2 and 4 show how aviation accidents can reach a level where they are perceived as crises necessitating change. The 1974 and 1979 DC10 crashes were the result of myopic policy decisions by both corporate and government elites. As the still dominant power in civil aviation, the United States was able to impose its safety standards on the rest of the world both before and immediately following the disasters.

Other regime participants felt vulnerable to the consequences of poorly implemented or poor policies. They had tired of having to abide by United States Federal Aviation Regulations. They felt that their standards more effectively achieved safety goals, and did not like paying the extra costs of compliance. They did not like bearing the costs of the 1979 DC10 grounding, especially when none of their DC10s had endured maintenance-induced structural damage. Since the 'definition of issues requiring regulation is a function of the perceived costs of asymmetrical interdependence' and 'calculations of sensitivity and vulnerability therefore inform the discussion of remedial measures' (Haas, 1980b: 363), those states which were able, namely European states, adopted policy options designed to bring them greater autonomy from perceived, real, and potential manipulation by the United States.

John Odell tells us, 'in any concrete situation, market pressure against an existing policy is always itself partly a reflection of previous action by governments as well as the international distribution of power among states' (1982: 28). A few years earlier, European elites had realised that technology, industry dynamics and market structure made it impossible for each state to build profitable competitive aircraft (see Chapters 3 and 4). They first attempted to develop their own certification standards when they decided to compete in the commercial airliner market as a group.[16] Now they pursued more vigorously the transnational production of Airbus aircraft and the development of standards via the Joint Airworthiness Requirements consortium. European Community members had learned to collaborate from experiments like the European Coal and Steel Community. They learned from mistakes, failures and successes encountered along the way that transnational management was possible.[17].

The loss of government financial support for US manufacturers sent producers abroad to gain (or keep) market share which would keep benefits flowing and to seek partners willing to share risk. The result was internationalised production strategies so successful that their redomestication is unlikely. Surplus capacity initially encouraged protectionism, but eventually impelled collaboration because public and private actors learned that more long-term gains were generated thus. It happened first in the manufacturing sector, probably because of 'the industry's degree of market concentration. The smaller the number of major producers, the greater the chance for co-operation in dealing with surplus capacity' (Cowhey and Long, 1983: 164; see also Olson, 1965; Strange, 1979; Tsoukalis and Ferreira, 1980: 358–60). With fewer participants at the bargaining table, aircraft manufacturers were able to organise collaborative projects.

Group size works well in explaining the otherwise remarkable lack of co-operation among the world's more than 300 airlines (ICAO, 1981: 52–7) in trying to match capacity with demands in the 1970s more efficiently. International airlines were unable to resolve the surplus capacity problem in spite of the presence of an international forum (IATA) specifically designed to assist in such matters.[18]

LOOKING BEYOND

Scholars, state and corporate elites and domestic constituencies now recognise that 'the national ability to disengage oneself from the network of economic interdependence is much less widespread than it was thought to be in the past' (Rosecrance, 1981: 699). Some fear the loss of autonomy associated with interdependence. The more sanguine observe that it impels co-operation and lowers the perceived possible gains associated with conflict. State and corporate elites have learned that co-operation can generate at least as many if not more benefits than conflict. Since 'successful negotiations for institutionalising international collaboration depend on the congruence of interests as much as on changes in consensual knowledge' (Haas, 1980b: 371), if state and corporate elites view international collaboration positively, we may see more of it.

Regimes have emerged and proliferated as international management tools for an increasing number of issue areas in our modern world. Understanding how actors create and shape regimes to fit their perceptions of how the world does and should operate helps uncover some of the puzzles of international relations. Accustomed to their creation and early maintenance by a hegemon in a less interdependent and more segmented anarchic society, the global community now asks, can states co-operate without the presence of a central authority? If so, which incentives, incorporated into regime principles, norms, rules and decision-making procedures, will impel co-operative behaviour so powerfully that states in a non-hegemonic system will not wish to defect?

This book suggests that a re-examination of sophisticated functionalist theory may be useful in seeking answers to these questions.[19] Neofunctionalists argue that in today's international system there are 'an increasing range of tasks and problems within the state, between them and transnationally' for which resolution will require action 'across national boundaries'. Further, the resolution of these problems is so vital that 'the problems of sovereignty' must be overcome (Taylor, 1975: 240–1).

Certification of Airbus aircraft is representative. The transnational consortium produces aircraft which must be certified by governments as safe before they can be marketed and sold. Separate state certification processes would increase costs and make the final price too high. Unwilling to yield sovereignty completely yet, participant state elites have worked out an informal procedure for accepting the authority of another state in the performance of this government function. These same elites are convinced a formally and legally binding means of sharing government authority will be established because of the economic imperative to achieve political co-operation (personal interview: Koplin, 1987; personal correspondence: Kennedy, 1981).

The dramatic failures of ill-conceived protectionist policies reflected in the 1974 and 1979 DC10 crashes add a negative sanction to these positive incentives urging co-operation across state boundaries.

Change is a process. It is evolutionary and generally unnoticed by participants until a crisis triggers awareness. When corporate elites pursue international production strategies, resources do transcend boundaries and are redistributed. These changes first occur at a very low level: the rules and decision-making procedures which once dictated national production followed by trade were superseded first by global 'outsourcing' – purchasing foreign-built components for final aircraft assembly domestically. Currently this process is being replaced by equity sharing, transnational, co-production processes. Similarly state or state-citizen controlled air transport services may be in the process of transition to transnationally produced services. Nation–state co-operation without a hegemon is possible and is happening on a regional scale already. What happens to the heretofore constitutive principles of state sovereignty and property rights remains to be seen.

Appendix 1 International Agreements Affecting Export Financing of Civil Aircraft

International agreements affecting the export financing of civil aircraft fall within the general framework of the Organisation for Economic Cooperation and Development (OECD) and the General Agreement on Tariffs and Trade (GATT). By 1982, there were three applicable OECD agreements:

1. *The Arrangement* (1976) sets rates and terms for export credits by country groups – rich, intermediate and poor. It was amended in November 1981 and again in June 1982. The Arrangement covers export credit terms for products other than civil aircraft and nuclear power equipment. Rates established in June 1982 are as follows:

	Number of years in repayment period		
Classification of Borrowing Country	*2–5*	*Over 5 to 8.5*	*Over 8.5 to 10*
Relatively rich	12.15%	12.4%	No credits
Intermediate	10.85%	11.35%	No credits
Relatively poor	10%	10%	10%

In addition there are special rates for countries graduating from 'poor' to intermediate.

2. *The Standstill (1974)*, as amended in May 1975, established down payment requirements, tenor of loans or leases and repayment terms for all sizes and types of civil aircraft. Export Credits for large commercial jet transports may cover no more than 90 per cent of the investment, may not exceed a 10– year tenor and must be repaid in no less than 20 equal semi–annual instalments of principle plus interest. The Standstill is silent on interest rates but the export credit agencies

have generally adhered to rates established in the Arrangement. The Standstill has no termination date but any signatory may withdraw upon giving the other signatories three months' notice.

3. *The Commonline (1981)* is a detailed subagreement to the Standstill and is applicable only to export financing in support of Airbus and large US manufactured civil jet transport sales to airlines in third party countries (export credit support for all other aircraft sales are still governed by the Standstill). Among other particulars, the Commonline established down payment requirements (minimum of 15 per cent), tenor of loan (maximum of 10 years), direct credit participation (maximum of 62.5 per cent of investment if credit repaid over entire loan period or 42.5 per cent of investment if credit repaid during last half of loan period) and interest rates (minimum of 12 per cent for US Dollars, 11.5 per cent for French Francs and 9.5 per cent for Deutsche Marks) (Beard, 1982: personal correspondence).

The GATT prohibits government subsidies of export financing if they lead to trade distortion. Moreover, any government that is a signatory of GATT may not extend export credit at rates lower than its average borrowing cost unless it is party to an OECD agreement which allows for such rate subsidies. Since the Arrangement excludes aircraft and Standstill is silent on the issue of interest rates, the only agreement that actually permits aircraft interest subsidies is the Commonline. It should be noted that both the European and American export credit agencies were consistently in violation of the letter of the GATT Subsidy Code from January 1980 through most of 1981 since government borrowing rates were higher than aircraft export credit agency rates. As of 1982 the export credit agencies supporting Fokker and BAC aircraft were technically still in violation of GATT since neither was a party to the Commonline (Beard, 1982: personal correspondence).

Appendix 2 Non-Tariff Barriers in the Airline and Aircraft Manufacturing Industries

Some examples of airline industry trade barriers are outlined by Betsy Gidwitz in *The Politics of International Air Transport* and by the Department of Commerce in *United States Service Industries in World Markets*:

1. *Excessive and Discriminatory User Charges* – these include airport and en-route navigation charges in excess of the cost of providing the services. Airport charges consist of landing fees with surcharges for night landing, parking and hangar usage; passenger charges; air navigation service fees, fuel nuisance charges, noise levies, and security charges. En-route navigation charges include fees for services provided by air traffic controllers, radio information and so forth.

2. *Ground Handling Requirements* – these include control over passenger check-in facilities, aircraft loading and towing, freight handling, maintenance and other ground operations. Governments often require foreign air carriers to use indigenous airport facilities, services and personnel, while giving preferential treatment of priority and performance to the national airline.

3. *International Currency Regulations* – these consist of restricting the use of local currency for purchasing tickets on foreign airlines; the imposition of heavy taxes on currency conversions; and the delay of foreign funds transfers earned by foreign carriers creating cash flow problems and potential financial losses from currency devaluations and lack of investment opportunities.

4. *Business and Other Taxes* – these are unevenly applied, to the detriment of the foreign air carrier.

5. *Cargo Restrictions* – these are similar to government procurement policies: government purchases of international air services are placed with national air carriers; freight handlers direct commercial cargo to the national airline; those tempted to fly their goods on

foreign airlines are 'encouraged' by government and lobby groups to change their plans.

6. *National Carrier Preferences* – these include requiring government officials to fly the national airline, and encouraging private citizens and companies via rebate, tax credits or imposition of travel restrictions to do the same.

7. *Marketing and Sales Restrictions* – these include limitations placed on the number of tickets, regardless of capacity, a foreign airline may sell, while adding benefits to the purchase of the national airline's tickets, such as better grade flight accommodations with room and board at certain stopovers; travel agents and agencies are encouraged via travel privileges and rebates to sell tickets on the national airline; and foreign airline advertising is restricted in those countries with nationalised media.

Non-tariff barriers within the aircraft manufacturing segment of the aviation industry are hard to identify and prove:

1. *Government Subsidy* – government ownership and financing of production often allows products to be sold at less than cost.

2. *Government Procurement* – parallels the 'national carrier preference' concept in the airline industry. Commercial, military and other government aviation equipment is often required or strongly encouraged to be of domestic origin.

3. *Standardisation or Safety Requirements* – discussed in the text.

Appendix 3 Regulation Creation Process

US Federal Aviation Regulations govern safety in civil aviation. They include certification standards for aircraft airworthiness, flight crew members, and airports, operational requirements for air carriers, and rules for air traffic control and air space utilisation. Rule making can be a complicated and time–consuming process; it is designed to eliminate interference of extraneous political and economic factors.

According to Ed Chapman, the 1981 Acting Chief of the FAA's Technical Standards Branch, there are three major motivations for the development of a new regulation or the modification of an old one: 1) A catastrophic air accident, investigated by the FAA and the National Transportation Safety Board, often results in regulation revisions to eliminate repeat occurrences. Aircraft airworthiness standards were modified within six months of the 1974 DC10 crash. 2) Research leading to changes in state-of-the-art aircraft technology can require regulation modification in order to maintain currency. 3) Special Conditions and Exemptions may be granted effectively modifying existing regulations. Special Conditions generally are applied to new generation aircraft because the advanced technology requires different treatment from that in old regulations. Exemptions are granted if a particular rule is not applicable to a given aircraft. In these cases the manufacturer must prove the exemption will not reduce aircraft safety and will benefit the public. This process can take as long as four to six years (1981: personal interview).

Initially a draft of the proposed regulation is submitted for agency consideration. If approved, it is brought to public attention as a Notice of Proposed Rule Making (NPRM) printed in the Federal Register; this is sometimes supplemented by a news release. The Notice describes the substance and form of the proposal and sets the time within which the public may submit written comments. During this public hearing, comments are received from the general public, interested organisations and specialised aviation industry groups. These comments and other supporting data are analysed and evaluated.

The FAA can decline to concur with any received information but it must provide a rationale for its position. The final draft of the rule is prepared and submitted to the FAA administrator for approval. If approved, it is incorporated into the existing Federal Aviation Regulations. The point to remember here is the amount of time involved in making rule changes. An emergency can take six months, and anything less takes years. Aircraft manufacturers and airline operators have learned to use circumventing power by acting first, then seeking exemptions or changes, otherwise it would bc virtually impossible for them to stay at the leading edge of their industry.

Appendix 4
The Certification Process

A three part certification process governs aircraft built in the United States: a Type Certificate (TC) is issued when it is determined an aircraft's design is airworthy. A Production Certificate (PC) is issued to manufacturers with quality control programmes insuring that aircraft are built according to the approved design and with high quality materials. An Airworthiness Certificate (AC) is issued when a final inspection of each aircraft built under both a TC and PC verifies its conformity to applicable regulations.

An Air Carrier Operating Certificate is issued to airlines which can ensure compliance with an aircraft's Airplane Flight Manual (AFM) and an FAA-approved maintenance programme.

The relationship between imponderables, industry dynamics and certification is important. To sell an aircraft, manufacturers make promises regarding performance, cost and delivery schedules. Once a contract is signed, these become legally binding commitments. If the manufacturer does not deliver, the airline can demand its 'pound of flesh' in payment. These commitments can be affected by any unforeseen difficulties necessitating changes or delays in the certification process. A primary manufacturer goal is to eliminate unpleasant surprises by helping establish safety standards and by co–ordinating certification and aircraft production schedules.

One of the most important parts of the certification process is the development of new Federal Aviation Regulations and 'Special Conditions' to evaluate the safety of new design features. Once the manufacturer applies for a Type Certificate, its engineers meet regularly with an FAA Type Certification Board (TCB), comprised of FAA aerospace engineers and flight test pilots. The TCB is responsible for analysing and monitoring design data submitted by manufacturer Designated Engineering Representatives (DERs). It also conducts and witnesses aircraft and component testing. Through this dialectic significant design changes can be made.

Few serious or significant design changes are demanded by the FAA. Manufacturer delivery commitments to airlines provide a partial explanation:

> In order to deliver aircraft as soon as possible after the Type Certificate is awarded, the manufacturer begins providing materials, 'tooling up' production lines, and subcontracting of components as much as 3 years before the TC is awarded. (United States House of Representatives, 1980: 15)

Once enough orders for a new aircraft have been placed to justify production, the manufacturer makes an Authorisation to Proceed (ATP) commitment and almost immediately begins more detailed design and component fabrication. It also applies for a Type Certificate, the awarding of which often takes 40–50 months. Before granting a TC, all 'paper analyses' must be reviewed and approved, and a prototype aircraft must be flight tested. By this time the manufacturer has made several design, tooling and subcontracting commitments. To require significant design changes could conceivably compromise aircraft integrity, reduce promised performance, increase guaranteed cost or slow down delivery schedules. A House Government Activities and Transportation Subcommittee concluded:

> economic pressures against FAA-encouraged or mandated improvements increase steadily as the manufacturer invests in an aircraft project, and accelerate suddenly – and perhaps irrevocably – after the manufacturer signs binding contracts with the air carriers and makes an Authorisation to Proceed (ATP) commitment. (United States House of Representatives, 1980c: 36).

Appendix 5 The Applegate Memorandum

27 June 1972
Subject: DC10 Future Accident Liability.

The potential for long-term Convair liability on the DC10 has caused me increasing concern for several reasons.

1. The fundamental safety of the cargo door latching system has been progressively degraded since the program began in 1968.
2. The airplane demonstrated an inherent susceptibility to catastrophic failure when exposed to explosive decompression of the cargo compartment in 1970 ground tests.
3. Douglas has taken an increasingly 'hard-line' with regards to the relative division of design responsibility between Douglas and Convair during change cost negotiations.
4. The growing 'consumerism' environment indicates increasing Convair exposure to accident liability claims in the years ahead.

Let me expand my thoughts in more detail. At the beginning of the DC10 program it was Douglas' declared intention to design the DC10 cargo doors and door latch systems much like the DC8s and 9s. Documentation in April 1968 said that they would be hydraulically operated. In October and November of 1968 they changed to electrical actuation which is fundamentally less positive.

At that time we discussed internally the wisdom of this change and recognized the degradation of safety. However, we also recognized that it was Douglas' prerogative to make such conceptual system design decisions whereas it was our responsibility as a sub-contractor to carry out the detail design within the framework of their decision. It never occurred to us at that point that Douglas would attempt to shift the responsibility for these kinds of conceptual system decisions to Convair as they appear to be now doing in our change negotiations, since we did not then nor at any later date have any voice in such decisions. The lines of authority and responsibility between Douglas and Convair engineering were clearly defined and understood by both of us at that time.

In July 1970 DC10 Number Two* was being pressure-tested in the 'hangar' by Douglas, on the second shift, without electrical power in the airplane. This meant that the electrically–powered cargo door actuators and latch position warning switches were inoperative. The 'green' second shift test crew manually cranked the latching system closed but failed to fully engage the latches on the forward door. They also failed to note that the external latch 'lock' position indicator showed that the latches were not fully engaged. Subsequently, when the increasing cabin pressure reached about 3 psi (pounds per square inch) the forward door blew open. The resulting explosive decompression failed the cabin floor downward rendering tail controls, plumbing, wiring, etc. which passed through the floor, inoperative. This inherent failure mode is catastrophic, since it results in the loss of control of the horizontal and vertical tail and the aft center engine. We informally studied and discussed with Douglas alternative corrective actions including blow out panels in the cabin floor which would provide a predictable cabin floor failure mode which would accommodate the 'explosive' loss of cargo compartment pressure without loss of tail surface and aft center engine control. It seemed to us then prudent that such a change was indicated since 'Murphy's Law'** being what it is, cargo doors will come open sometime during the twenty years of use ahead for the DC10.

Douglas concurrently studied alternative corrective actions, in-house, and made a unilateral decision to incorporate vent doors in the cargo doors. *This 'bandaid fix' not only failed to correct the inherent DC10 catastrophic failure mode of cabin floor collapse, but the detail design of the vent door change further degraded the safety of the original door latch system* by replacing the direct, short-coupled and stiff latch 'lock' indicator system with a complex and relatively flexible linkage. (This change was accomplished entirely by Douglas with the exception of the assistance of one Convair engineer who was sent to Long Beach at their request to help their vent door system design team.)

This progressive degradation of the fundamental safety of the cargo door latch system since 1968 has exposed us to increasing liability claims. On 12 June 1972 in Detroit, the cargo door latch electrical actuator system in DC10 number 5 failed to fully engage the latches of the left rear cargo door and the complex and relatively flexible latch 'lock' system failed to make it impossible to close the vent door. When the door blew open before the DC10 reached 12 000 feet altitude the cabin floor collapsed disabling most of the control to the tail surfaces and aft center engine. It is only chance that the airplane was not lost. Douglas has again studied alternative corrective actions and appears to be applying more 'band-aids.' So far they have directed us to instal small one-inch diameter, transparent inspection windows through which you can view latch 'lock-pin' position, they are revising the rigging instructions to increase 'lock-pin' engagement and they plan to reinforce and stiffen the flexible linkage.

It might well be asked why not make the cargo door latch system really 'fool-proof' and leave the cabin floor alone. Assuming it is possible to make the latch 'fool-proof' this doesn't solve the fundamental deficiency in the airplane. A cargo compartment can experience explosive decompression from a number of causes such as: sabotage, mid-air collision, explosion of combustibles in the compartment and perhaps others, any one of which may result in damage which would not be fatal to the DC10 were it not for the tendency of the cabin floor to collapse. The responsibility for primary damage from these kinds of causes would clearly not be our responsibility. However, we might very well be held responsible for the secondary damage, that is the floor collapse which could cause the loss of the aircraft. It might be asked why we did not originally detail design the cabin floor to withstand the loads of cargo compartment explosive decompression or design blow out panels in the cabin floors to fail in a safe and predictable way.

I can only say that our contract with Douglas provided that Douglas would furnish all design criteria and loads (which in fact they did) and that we would design to satisfy these design criteria and loads (which in fact we did).*** There is nothing in our experience history which would have led us to expect that the DC10 cabin floor would be inherently susceptible to catastrophic failure when exposed to explosive decompression of the cargo compartment, and I must presume that there is nothing in Douglas's experience history which would have led them to expect that the airplane would have this inherent characteristic or they would have provided for this in their loads and criteria which they furnished to us.

My only criticism of Douglas in this regard is that once this inherent weakness was demonstrated by the July 1970 test failure, they did not take immediate steps to correct it. It seems to me inevitable that, in the twenty years ahead of us, DC10 cargo doors will come open and I would expect this to usually result in the loss of the airplane. This fundamental failure mode has been discussed in the past and is being discussed again in the bowels of both the Douglas and Convair organizations. It appears however that Douglas is waiting and hoping for government direction or regulations in the hope of passing costs on to us or their customers.

If you can judge from Douglas's position during ongoing contract change negotiations they may feel that any liability incurred in the meantime for loss of life, property and equipment may be legally passed on to us.

It is recommended that overtures be made at the highest management level to persuade Douglas to immediately make a decision to incorporate changes in the DC10 which will correct the fundamental cabin floor catastrophic failure mode. Correction will take a good bit of time, hopefully there is time before the National Transportation Safety Board (NTSB) or the FAA ground the airplane which would have disastrous effects upon sales and production both near and long term. This corrective action becomes more expensive than the cost of damages resulting from the loss of one planeload of people.

F. D. Applegate
Director of Product Engineering

* We [Eddy, Potter and Page] have been unable to establish whether Applegate's reference to an accident involving Ship Two, in July 1970, is a mistake on his part or whether there were two blowout incidents. Certainly Ship One was damaged on May 29, 1970, in circumstances very similar to those described by Applegate.

** Murphy's Law: 'If it can happen, it will.' Also known sometimes as the totalitarian or Hegelian law of physics, from Hegel's view that all that is rational is real: that is, if you can think of it, it must exist. This is a case where implausible philosophy makes for good engineering.

*** Douglas's design criteria called for the floor to withstand a pressure of 3 psi, and it eventually did so – although not until Douglas had challenged Convair's original stress analysis, and a stronger kind of aluminium alloy had been introduced to the floor beams (*Source for appendix*: Eddy *et al.*, 1976: 183-5; first emphasis mine, second emphasis and footnotes theirs).

Appendix 6 Government Liability Issues

In 1978, two FAA lawyers published an article in the *Journal of Air Law and Commerce* in which they claimed the government should not be held liable for damages caused by negligence in its performance of aircraft certification. John R. Harrison and Phillip J. Kolczynski argued that the 1958 Federal Aviation Act imposes regulatory responsibilities 'designed to secure the safety of the general public instead of establishing a legal duty of care to individuals'; therefore, 'there can be no cause of action in tort against the United States'. Second, the 'discretionary function exception' to the Federal Tort Claims Act 'immunises' the government against a lawsuit even when the FAA 'fails to detect a manufacturing or design defect during the certification process'. Third, an Airworthiness Certificate issued to an aircraft, 'which in retrospect proves to be unairworthy, is nothing more than misrepresentation', a mistake exempted from lawsuit by Congress (US Senate, 1978: 42-3).

They further argued that Constitutional separation of powers requires the courts not to 'arrogate to themselves the [FAA's] adjudicative function'. Safety cannot be mandated. It is impossible to establish regulatory standards stringent enough to prevent, absolutely, accidents. And the cost of achieving perfect safety would economically paralyse the aviation industry. 'Such over-regulation would be contrary to the Administrator's responsibility to foster the development of civil aeronautics' (1978: 43).

Harrison and Kolczynski also contended that the courts should 'consider the economic policy implications of imposing liability on the United States for the erroneous certification of aircraft' (1978: 43-4). If a 'manufacturer launches a defective product into the stream of commerce which creates an unreasonable risk of harm', it will be held liable for the injury and damage costs caused by the defect.

> The principle of commutative justice assists the manufacturer by allowing the burden to be spread among the consumers who enjoy the benefits of air travel. To impose liability on the United States for alleged errors in certification distorts the commutative system

> of justice, which contemplates spreading the loss once, primarily to the manufacturer, and then only to the flying public, not to the general public. (1978: 44-5)

The US Supreme Court has accepted this argument. On 19 June 1984, the Court ruled the FAA 'cannot be sued for carelessly failing to find a defect that causes a commercial air disaster' (Epstein, 1984: 15A). According to Chief Justice Warren E. Burger, who wrote for a unanimous Court, 'The FAA has a statutory duty to promote safety, not insure it' (as quoted in Epstein, 1984: 15A). The primary responsibility for air safety belongs to manufacturers and airlines; the FAA is required to encourage compliance with safety standards. A contrary decision

> would have exposed the taxpayers to liability in many of the 300 current air disaster claims exceeding $3 billion in damages . . . if the FAA were held legally responsible for negligent inspections, the government would become subject to damage suits for negligent inspections of . . . virtually all other federal programs to protect health and safety. (Epstein, 1984: 15A)

The US government is shielded by federal law 'from claims based upon a "failure to exercise or perform a discretionary function or duty." ' Since the FAA selection of a 'spot-checking method' was one among many options, chosen because it best accommodated its safety goal, given its limited resources, 'the courts should not "second guess" that judgement'; it is 'a "discretionary function" and thus immune from suit' (Epstein, 1984: 15A).

Notes and References

1 Introduction: Power, Process and Crisis

1. The engine is designed to separate this way since a red-hot engine hurtling forward is less likely to ignite fuel in the plane's wing tanks.
2. A sectoral analysis is 'any study of the political economy of a specific industry in its world context, or of specific markets for goods and services' (Strange and Tooze, 1981: 12).
3. Charles Perrow reminds us of an important and often neglected caveat to these statistics. 'Safety comparisons with other systems are hard to make. In many respects, commercial air travel appears to be much safer than automobile or rail travel. Many fewer people are killed in the first than in the other two. But an equally useful statistic would be the number of fatalities per hour of exposure, or per million miles traveled. Unfortunately, we do not have these statistics for automobile accidents . . . if we used the statistic of fatalities per 100 000 hours of exposure, highway travel would be the safest mode of transportation . . . One's chances of being killed while driving a car is only one per cent in fifty years of driving. We simply do a lot of driving and very little flying, giving us the impression that the risk of the latter is much smaller' (1984: 126; see also Golich, 1988 and Lederer, 1982).
   John O'Brien, Director of Safety for the Air Line Pilots' Association, and the National Academy of Sciences Transportation Research Board agree the level of safety intrinsic to any means of transportation is related to the issue of *risk exposure* not absolute number of catastrophes (O'Brien, 1987; Transportation Research Board, 1988).
4. This analysis is based on the assumption that there is an ideal level of safety which is unattainable no matter how well-intentioned and rigorous the system. It accepts Perrow's argument that 'Since nothing is perfect . . . there will be failures . . . [which will] become accidents rather than incidents because of the nature of the system itself; they are system accidents, and are inevitable, or 'normal', for these accidents' (1984: 330; see also Lederer, 1982).
5. I wish to emphasise this should in no way influence the reader to think other nations have not been or do not want to be involved in the aviation safety regime. For a discussion of the Third World's relationship to and participation in the international commercial aviation regime, see Stephen Krasner's *Structural Conflict: The Third World Against Global Liberalism* (Berkeley: University of California Press, 1985), especially Chapter 7.
6. For a review of regime theory literature, see Haggard and Simmons, 1987: 491–517.
7. Ikenberry is by no means the only scholar who has recognised this fundamental aspect of state behaviour or the need to examine the relationship between domestic policy processes and the structure

of the global environment. For further elaboration see Gowa, 1985, 1987; Jönsson, 1986, 1987; Keohane, 1986: 9, 19, 25; McKeown, 1986; Rosenau, 1986: 852–859, 890+; Russett, 1985: 227; Strange, 1985: 237; and Willetts, 1982.

8. See Hart, 1986: 19; Kratochwill and Ruggie, 1986: 767–51; Rosenau, 1986: 852–53, 890+; Strange, 1982 and 1985; Strange and Tooze, 1981.
9. See Biersteker, 1987; Haas, 1980; McKeown, 1986; Odell, 1982; Rosenau, 1986; Smith, 1987.
10. Effective behaviour leads to the achievement of stated goals; those goals may turn out to be myopic or poorly conceived. Nevertheless, it is important to know which policies led to their achievement.
11. For more on how advancing technology might trigger regime change, see Haas and Ruggie, 1976. The role of ideas and knowledge is analysed by Haas, 1982, 1980; Odell, 1982. Complacency leading to a crisis that acts as a catalyst to change is discussed by Krasner, 1976. Goldstein, 1986 and Lindblom, 1981 address other domestic factors which could combine with any of the above variables to cause states to shift their adjustment strategies.

2 The Evolving Safety Regime

1. See Chandler and Daems (1980) for a discussion of how large transportation systems such as railroads and airlines require centralised control and co-ordination to be efficient.
2. The importance of range was underscored by the outbreak of the Second World War, when the Allies needed to span the North Atlantic and leapfrog hostile countries to maintain political ties with the scattered empires. In the 1930s, Great Britain had focused on flying boats, and the Germans had shifted their attention to producing a good bomber. Aided by its geography amd Second World War Allied responsibility, American manufacturers concentrated on building longer-range transport aircraft. Soon the airlines of the smaller European countries (Holland's KLM, Switzerland's Swissair, Belgium's SABENA), which were not committed to purchasing home-produced aircraft since they did not build their own, began to order US aircraft.
3. Territory was defined as 'including the national territory, both that of the mother country and of the colonies, and the territorial waters adjacent thereto' (as quoted in Salacuse, 1980: 813).
4. The US attended the Paris Conference and signed the Convention text, but it never signed the treaty. Policy makers were uncomfortable with ICAN's power to issue legally-binding rules and with its link to the League of Nations (Salacuse, 1980: 813fn25, 831fn115).
5. This was known as the Morrow Board after its Chair, Dwight D. Morrow.
6. This impelled the technological transition to all metal aircraft construction.

7. In 1923, the US Air Mail Service, still operating as a government entity, won the Collier Trophy for operating 12 months without a fatal accident, although 'one in every six pilots was killed before and after' (Lederer, 1987: personal correspondence).
8. It is not easy to predict what is publicly perceived to be an unacceptable accident rate.
9. Of the 55 allied and neutral states invited to Chicago, 1 November 1944, 52 attended (Gidwitz, 1980: 47):

Afghanistan	France	Panama
Australia	Greece	Paraguay
Belgium	Guatemala	Peru
Bolivia	Haiti	Phillippine Commonwealth
Brazil	Honduras	Poland
Canada	Iceland	Portugal
Chile	India	Spain
China	Iran	Sweden
Colombia	Iraq	Switzerland
Costa Rica	Ireland	Syria
Cuba	Lebanon	Thailand
Czechoslovakia	Liberia	Turkey
Denmark	Luxembourg	Union of South Africa
Dominican Republic	Mexico	United Kingdom
Ecuador	Netherlands	United States
Egypt	New Zealand	Uruguay
El Salvador	Nicaragua	Venezuela
Ethiopia	Norway	Yugoslavia

10. Robert Thornton's table offers a hypothetical example of the problem (1970: 31):

Effect of Failure to grant fifth-freedom rights on outbound Pan American Flight 2: New York–New Delhi–around the world

City	*Persons deplaning*[a]	*Persons continuing*	*Outgoing load factor (%)*
New York	—	100	67[b]
London	60	40	27
Frankfurt	31	9	6
Istanbul	2	7	5
Beirut	4	3	2
Teheran	2	1	1
New Delhi	1	0	—

[a] Assumptions about deplaning are based on Pan American Airways' usual percentage between New York and terminals listed, as provided by the US Civil Aeronautics Board, 'International Origin–Destination Survey of Airline Passenger Traffic 1961' (Washington, DC: Civil Aeronautics Board).
[b] Aircraft capacity is assumed to be 150.

11. A proposal granting all five freedoms – the International Air Transport Agreement – was drafted, but so few states signed it that it was dropped in 1946.
12. Annex 8: International Standards – Airworthiness of Aircraft, adopted 1 March 1949, was ICAO's first attempt at establishing an international airworthiness code. It includes broad standards which define, for application by national airworthiness authorities, the minimum basis for recognising Certificates of Airworthiness which would allow foreign aircraft to penetrate a state's air space. One of the weaknesses of Annex 8, however, was that it prescribed standards for only one category of aircraft. In 1957, an attempt was made to rectify this situation via the adoption of Amendment 85 to Annex 8. As a result, the ICAO Council no longer attempts to establish airworthiness standards for various categories of aircraft; instead it has adopted a more general airworthiness policy which establishes principles and goals to be contained in the comprehensive airworthiness codes of Contracting States (for further discussion see Buergenthal, 1969: 86–8).
13. States recognised early that political and economic gains accompanied the adoption of standards for manufacturing. The country whose company held the patent or appropriate standards could generate valuable income while other companies had to invest in retooling their factories to compete.
14. A Bilateral Airworthiness Agreement (BAA) regulates the quality of aeronautical equipment for import or export. The purpose is to guarantee that a traded aeronautical product is airworthy. Extraneous issues affect these agreements but their primary concern is with safety. These Agreements exist only among the 24 countries which produce aircraft or aircraft components. Among the first to be concluded were those between the US and Canada in 1929, the US and Italy in 1931, and the US and the UK in 1934.
15. These are called Bilateral Air Transport Agreements, and they govern the air transport relations between the two signatories. They regulate the awarding of routes, landing rights, capacity levels, fare-setting, frequencies, gateways, and so forth. They often specify security arrangements designed to prevent aerial hijackings. Virtually every participant in the world's air transport system is a signatory to many Bilateral Air Transport Agreements, as a separate one must be concluded with every country to which a state's airline(s) flies.
16. The Department of Transportation was created in 1966.
17. The NTSB can only recommend procedural or technical changes Board members believe would increase safety by preventing similarly caused accidents. The FAA decides which will become legally-binding regulations.
18. Here I am using Stephen Krasner's definition of regime and its elements as found in *Structural Conflict*: 'Regimes are principles, norms, rules, and decision making procedures around which actor expectations converge. Principles are a coherent set of theoretical statements about how the world works. Norms specify general standards of behaviour. Rules and decision-making procedures refer

to specific prescriptions for behaviour in clearly defined areas' (1985: 4).

19. A collective goods issue area is one which would benefit everyone if properly managed and perceived of as capable of distributing mutual benefits to all who participate in it. Unfortunately, the cost–benefit equation for a variable sum management scheme is seldom clear enough to compel effective co-operation among its participants.
20. For a more extensive discussion of the roles of technological advances and knowledge see Chapter 5; Haas, 1982: 23–59; 1980b: 367–8; Kratochwil and Ruggie, 1986: 773; Rosenau, 1986: 864; Smith, 1987: 278–81.
21. This leaves open some intriguing questions regarding how to regulate outer space vehicles, satellites, etc.

3 The Value and Dynamics of Civil Aviation

1. This could have the positive effect of impelling state co-operation in arenas otherwise the domain of unilateral, zero sum game type behaviour. Klaus Koplin, Chief of the Engineering Division for West Germany's civil aviation agency and responsible for working out certification processes and standards with his colleagues in those countries participating in the production of Airbus commercial aircraft, contends that corporate responses to a global market have forced state level co-operation in Europe (1987: personal interview). This topic is explored in a paper presented at the North Eastern International Studies Association (Golich, 1987).
2. The Standard Industrial Classification (SIC), a system developed by the US government to define the economy's industrial composition, facilitates statistical comparability. It is revised periodically to reflect relevant changes, and the definitions are published by the Office of Management and Budget. The figures used in the following discussion reference the Aircraft and Parts (SIC 372) and Guided Missiles and Space Vehicles (SIC 376) parts of the industry. The latter category is not the subject of this study and is only mentioned where necessary. The reader should note that where the term 'aerospace' is used, it refers to SIC 372 and SIC 376 combined (for more detailed definitions see Bluestone *et al.*, 1981: 5; and *Aerospace Facts & Figures 1986/87*, 1986: 12).
3. These are figures for scheduled airlines only.
4. These figures exclude Spain and Portugal for which data was not available.
5. Turnover is a term used to denote production net of change in backlog; it is roughly equivalent to sales.
6. Technology-intensive industries are defined by the intensity of the new-technology-generating inputs used in their production – research and development, scientific and engineering manpower used in functions other than R & D (design, production supervision, customer services, etc.) and the relative level of the worker's skill (Boretsky, 1975: 74).

7. Stephanie Neuman argues that military production hierarchy in the Third World follows a different progression. It starts with naval craft, then aircraft, missiles and armoured vehicles, in that order (1983: 175). This discrepancy is most likely explained by the degree of sophistication of the aircraft produced by these developing countries. She, in fact, notes that the aircraft to which she refers are 'simple trainers', and even they are 'increasingly beyond their [LDC] need and capability'. Bacher, on the other hand, has identified more sophisticated types of military aircraft such as logistical and fighter aircraft.
8. Robert Gilpin cautions that the concept of prestige defies measurement and is difficult to define. Here it is understood to mean the reputation for power. Prestige rests on economic and military capabilities which have been successfully used to achieve state objectives. It rests on 'the perceptions of states with respect to a state's capacities and its ability and willingness to exercise its power' (1981: 29–33). States which enjoy a reputation for prestige are fortunate because 'if your strength is recognised, you generally achieve your aims without having to use it' (E.H. Carr, as quoted in Gilpin, 1981: 31). For a detailed discussion of the concept of prestige see Gilpin (1981: 28–34).
9. Ever since Woodrow Wilson created the National Advisory Committee for Aeronautics in 1915, to review and co-ordinate all aeronautical research and technology policies, virtually every president has appointed similar advisory groups. Gerald Ford's steering committee, charged with developing policies to foster a healthier aviation industry, was the first to suggest economic deregulation of airlines. This policy was successfully pursued by Jimmy Carter.

   Originally, Ronald Reagan wanted to include aeronautical research in his budget-cutting agenda. It was generally felt that 'after years of public support, we might be at a point of diminishing returns' (George A. Keyworth as quoted in Coleman, 1982: 22). A steering committee and working group were created to determine how and where to reduce aerospace spending; instead, they concluded aeronautics still had 'immense strategic value – both for defense needs and for the economic benefits'. In the service sector, aircraft were found to be 'the dominant common carrier for intercity travel', underscoring federal responsibility to maintain safety in the air transport system (Coleman, 1982: 22).
10. From 1984 to 1987, European Community transport ministers negotiated an economic liberalisation package. The deal was championed by the United Kingdom and the Netherlands who had already begun to economically deregulate their mutual skies. The agreement gives airlines more freedom to offer discounts of as much as 65 per cent off standard fares and makes it easier to start new flights between desirable locations. The decision is a double-edged sword: on the one hand, European carriers want a wider European market to help them become stronger against American competitors. 'If the EC does succeed in creating one set of rules for a single air market by its 1992 deadline, then it may want to negotiate a pan-European air pact with the US to replace a web of bilateral agreements' (Heard, 1987: 54). On the other hand, with only 320 million consumers, Europe is not

big enough to keep 21 national airlines. Issues of national sovereignty will probably delay the rationalisation process in Europe, though the eventual result will no doubt be transnational 'megacarriers' within the community.

11. Sloppy record-keeping makes it difficult to identify which companies use which government facilities. According to Robert J. Gordon of Northwestern University,

    > the aircraft industry is a virtual ward of the government, having contributed approximately 10% of the funds for its own expansion in World War II and only about one-third during the Korean War . . . Much of the aircraft and ordnance production for the Vietnam War [was] carried out by private firms with government-owned plant and equipment. (Gordon, 1969: 224).

12. An Air Force purchase of 50 modernised Lockheed C–5B airlifters and 44 McDonnell Douglas KC10 tankers (a military version of the DC10), would have boosted the earnings of both companies substantially and extended the DC10 production run. The latter was important because DC10 orders had dropped so drastically by December 1982 that production was terminated. Nevertheless, Boeing appeared to be most desperate; profits had plummeted 38 per cent and nearly $5 billion in orders for its 747s and newer 757s had been cancelled ('A Bitter Clash . . . ', 1982: 91–2).
13. This is a more complex issue than it appears on the surface. Generally, US airlines must purchase domestically-produced aircraft with loans from commercial banks, which require sufficient traffic potential and ready cash or capital assets to justify such an expenditure. The US Export Import Bank offers low interest loans with less stringent conditions to foreign carriers to encourage exports. Then, once the loan has been extended to a foreign carrier, the selling government has an interest in that carrier's success and may be more generous in granting landing rights to protect its investment (Kalijarvi, 1957: 1009, 1013).

    All states have institutions which perform functions similar to those of the US Exim Bank. These are *Hermes* in the Federal Republic of Germany, *Compagnie Française d'Assurance du Commerce Extérieur* in France, and the Export Credit Guaranty Department in the United Kingdom. The US also has the Private Export Funding Corporation, a 'private corporation owned by 54 commercial banks, seven industrial corporations and one investment banking firm. It lends to foreign private and public borrowers seeking medium- and long-term financing to buy US goods and services. All of its loans must be covered by an unconditional guarantee of Exim Bank'. (Mann, 1983: 197; see also Newhouse, 1982: 59–60).
14. The agreement raised the minimum interest rate charged for government loans to 12 per cent for the first 62.5 per cent of the aircraft sale price and limited the loan terms to 10 years. Prior to this the three European governments had been offering interest rates of 8.75 per cent on aircraft export loans, while the US Exim Bank's rate

was 9.25 per cent. (For more detail regarding international agreements which affect the export financing of civil aircraft see Appendix 1).

15. The 1977 events surrounding the admittance of the Concorde supersonic jet into US airspace demonstrate how standards can be used to achieve political goals. Development of a US supersonic jet was postponed due to conservationist objections about noise levels, while the British and the French continued their production of the Concorde and sought an agreement to allow its use on trans-Atlantic routes. Objections to the noise problems eventually led to congressional hearings. The administration feared anti-Americanism would be aroused by US denial of Concorde entry rights, and this might aid the election of a socialist–communist coalition in France. Therefore, National Security Adviser, Zbigniew Brzezinski, instructed two Department of Transportation officials not to testify. Nevertheless, Concorde gateways and air space which could be penetrated at supersonic speeds were restricted.
16. As of December 1987, this included Continental, Eastern, New York Air and People Express.
17. New generation aircraft incorporate new technology responsive to economic pressures. The first equipment cycle began with the advent of the commercial jet transport in the late 1950s in response to a demand for long distance, fast, comfortable transportation. The second encompassed the development, production and sale of 'jumbo jets' in order to increase passenger mile capacity while decreasing the number of overcrowded airplanes. 'Environmental pressures and the "energy crisis" are responsible for the latest generation of commercial transports' (Bluestone *et al.*, 1981: 47).
18. The Air Transportation International Research Forum, industry forecasters, and the Federal Aviation Administration all try 5-10-year projections but feel more comfortable with 5-year predictions.
19. When airlines buy new equipment, they incur costs in three areas other than the capital acquisition. Personnel must be trained; new facilities must be built; and a spare parts inventory must be purchased and stored. These activities are timed to coincide with the aircraft arrival. If they are completed too late, the airline will incur the costs of an idle aircraft parked on the apron; if they are completed too early, the airline will incur the costs of unusable manpower and facilities. Airlines make their operational plans in expectation of utilising the new equipment. They generally try to introduce a new aircraft during the winter so any problems can be worked out before the summer boom. Airlines whose new purchases arrive late sometimes sue the manufacturer for the revenue losses suffered because of the delay. In the mid-1960s, McDonnell Douglas was having difficulty meeting its DC9 delivery dates; Eastern Airlines sued for 'inexcusable delays' which had wreaked havoc with its planned schedules (Eddy *et al.*, 1976: 70).
20. United Nations' figures measure the volume of world trade in aircraft and parts. They do not include domestically produced and sold aerospace products, nor do they include missiles and space products included in US trade figures. World aircraft production

figures and trade in missiles and space products are not readily available.

21. For purposes of this study, civil aircraft and commercial class aircraft are used interchangeably. They refer to a class of aircraft defined by the FAA as over 12 500 takeoff weight.
22. A revenue passenger mile represents the number of paying passengers who fly one mile.
23. Even though historically price has not been a determining factor in aerospace sales, the more lenient the terms, the more likely the sale. In 1978, Airbus Industries was able to offer Eastern a deal too good to refuse ('Europeans Press . . . ', 1978: 51-3; Griffiths, 1978: 27–8; Kozicharow, 1978: 59; Carley, 1978: 14; 'Eastern Air . . . ', 1978: 2). The entire package amounted to $778 million for 23 aircraft to be delivered over a 5-year time span. Airbus arranged external financing for a total of $693 million, including guarantees by participating state export credit agencies. In addition, the deal included several 'sweeteners' not normally included in aircraft sales contracts: a 6-month no-cost lease for the initial four aircraft for evaluation purposes; Airbus Industries paid all legal fees, US tariffs, FAA certification costs and maintenance and repair costs associated with introducing the aircraft into Eastern's fleet. Finally, because the plane was not considered by Eastern management to be the optimum size for the airline to approach economic efficiency on the basis of seat-cost-per-mile, Airbus offered to pay certain operating costs as a bridge to the availability of a smaller version (150-seat) transport. In testimony before the House Subcommittee on International Trade, J.B.L. Pierce, Boeing's Treasurer, noted, 'We can compete with Airbus and other European aircraft manufacturers on cost and technical merits, but we cannot compete with the national treasuries of France and Germany' (as quoted in Johnsen, 1978b: 59).

    McDonnell Douglas countered with a new short-term lease concept. Leasing arrangements usually span an 18-year period. DC9–80 aircraft were offered to airlines for 5 years, after which they could be returned to the manufacturer without any penalty, or the agreement could be extended for 13 years. Both supplier and consumer benefited: the latter gained access to new generation equipment with minimal effect on its borrowing capacity and without the risk of soon being burdened with obsolete aircraft. The manufacturer benefited because production runs and work force levels were stabilised, lowering aircraft production costs (Smith, 1982: 28–9; Williams, 1983: personal interview).
24. This may be a short-term rather than a structural problem (Willy, 1986: personal interview).
25. Experts now predict aircraft can last 30, 40 and 50 years as long as they are properly maintained. Once, they were assumed to have only a 20-year service life (see *The Competitive . . .* , 1985: 36). At 25 years, they were considered antiques.
26. Even if fuel costs do not increase dramatically, utilising fuel-efficient equipment could provide a windfall for airlines if the savings are not passed on to the consumer by means of further fare discounting: 'A 1 cent per gallon drop translates to about $100 million in cash flow

for US carriers, and US airlines use about 10 billion gallons per year' (O'Lone, 1983: 169).

27. The Airbus Consortium consists of France's Aerospatiale which builds the nose and flight deck section, does some airframe work and handles assembly line duties in Toulouse, the United Kingdom's British Aerospace which constructs the wing, West Germany's Messerschmitt-Boelkow-Blohm (MBB) via the Deutsche Airbus organisation which manufactures portions of the fuselage, Spain's CASA (Construcciones Aeronauticas S.A.) which makes the horizontal stabiliser, and some small manufacturing done by the Netherlands' Fokker and Belgium's Belairbus.
28. It is important to recognise market access potential may be improved, but not guaranteed. For example, the Boeing 767 incorporates major work from Italy, but has not yet penetrated that market (O'Lone, 1983: 225).
29. See *The Competitive Status of the US Civil Aviation Manufacturing Industry* (1985: 61–6) for a more detailed discussion of this issue.

4 Disaster in the Skies

1. It is important to note regimes do not have to be characterised by hegemonic stability as implied here. Several studies convincingly argue that most were created and initially maintained by a hegemon. See for example, Keohane and Nye, 1977; Kindleberger, 1973; Gilpin, 1975; Young, 1982; and Avery and Rapkin, eds 1982.
2. See for example, Axelrod, 1985; Keohane, 1984; and Krasner, 1982.
3. Norms and principles may also change, of course, but not as easily.
4. Robert Gilpin may be the strongest proponent of this argument which asserts that the redistribution of power and wealth away from the hegemon negatively affects its ability to achieve goals. It is important to differentiate between absolute power and relative power in this discussion. It is also important to distinguish between power held and exercised in various issue areas.
5. For a thorough discussion of this issue see Golich, 1988; Lederer, 1982; O'Brien, 1987; Perrow, 1984.
6. This does not make it easier to identify when crises will occur or even what might be identified as crises in some less sensational manner or fields of inquiry.
7. See Appendix 3 for a detailed description of the regulation creation process.
8. In 1982, an Air Florida plane crashed into the Potomac River. Although the seats were fastened to the floor according to FAA standards, they were not as securely fastened as they could have been, adding to the impact forces and causing a majority of deaths. Subsequently, two House Science and Technology subcommittees – Transportation, Aviation and Materials, and Investigations and Oversight – conducted a joint investigation regarding the applicability of present technology

to the improvement of air safety. In the final report they concurred with a National Transportation Safety Board conclusion that a key factor in air accident survivability 'is the potentially lethal nature of the aircraft interior . . . and the inability of the seats and floor connections to restrain passengers in a crash' (1982: 27).

9. Among the imponderables are the wide range of unknowns John Newhouse discusses in *The Sporty Game*: 'known unknowns' and 'unknown unknowns', referred to as 'unk-unks'.

> Known unknowns are the normal, unremarkable improvements that, it is assumed, will be called for sooner or later; the problems and added costs such unknowns create are familiar and in a general way expected. Unk-unks are the less predictable contingencies; the assumption is that any new airplane or engine intended to advance the state of the art will harbor its own surprises in the form of problems that are wholly unforeseen . . . . Sometimes, as in the case of the metal fatigue which ended the Comet program, an unk-unk is not discovered until crashes have occurred. (1982: 19)

10. There is no intention to imply that any individual or entity in the aviation industry maliciously attempts to reduce air transport safety. No aircraft manufacturer would cut corners in its production process and knowingly compromise safety. Karl G. Harr, Jr., former President of the Aerospace Industries Association, Inc., in testimony before a House Subcommittee investigating the FAA's certification process, explained:

> The actual and perceived safety of the . . . air transport system is of crucial importance to us as manufacturers. Our livelihood depends on it. Not one of our planes would enjoy a successful market if it were not as safe as humanly possible. (United States House of Representatives, 1978: 264)

Marketplace failures of the Comet and Electra support this argument. The public lost confidence in these planes when they mysteriously disintegrated in midair and neither enjoyed a successful production run. One of the reasons McDonnell Douglas' DC10 initially enjoyed greater market penetration than Lockheed's L1011 was because of the reputations each had earned: Douglas aircraft were known to be conservatively built and extraordinarily reliable, whereas Lockheed's Electras had fallen out of the skies.

Some airline operations are safer than others also (Federal Aviation Administration, 1984; Ott, 1988: 98–101). The inequality in safety records is overlooked in an attempt to reassure the public that flying is the safest mode of travel available:

> Less safe airlines could not operate without the acquiescence of the entire aviation community: the insurers, the international bodies that are supposed to regulate and monitor the industry, and the airlines and their own industrial organization, who have done more than anyone to promote the myth of universal excellence. (Eddy *et al.*, 1976: 190)

Governments co-operate with the creation and preservation of the notion that air travel is the safest transportation because it is useful as a bargaining tool in obtaining 'commercially vital landing rights' (Eddy *et al.*, 1976: 206). But should there be successive accidents due to carrier malfeasance and the public knows about it, surely alternate carriers or other travel modes would be chosen. Aircraft manufacturers and airline operators attempt to provide safe air transport because of the economic benefits to be reaped, but also because no one would wish to be responsible for catastrophic human tragedy.

11. In 1979, Representative Barry M. Goldwater, Jr., figured it would cost the government $90 million a year to hire FAA engineers and inspectors to replace Designated Engineering Representatives (DERs). Even then it would be hard to 'lure' qualified persons away from private industry. Jake Karnowski, former Assistant Engineering Chief for the FAA Western Region, explains: 'We can't compete. We can't offer the money that private industry can' (Green *et al.*, 1979: 1+). Some argue that using DERs benefits the process by giving the FAA access to many of the best aviation engineers.
12. Generally 'conflict of interest' is not considered a serious problem: DER appointments are made annually on the basis of a good performance record. In its study of the certification process, the National Academy of Sciences defended the DER system claiming the potential for conflict is checked in the following ways:

    (i) engineers are ethically motivated to maintain their reputation for technical integrity and professionalism;
    (ii) recognising the stake of the manufacturer in assuring a safe, serviceable, and reliable airplane, the company's designees perform traditional engineering review tasks for the FAA that would . . . be performed for the company as well;
    (iii) the designees perform their work under the supervision of the FAA staff; and
    (iv) the FAA reserves to its own staff the most critical design decisions and approvals (Low, 1980: 7).

13. For example, part of the DC10 door design which proved to be a contributing cause of the 1974 disaster was submitted to a series of tests by McDonnell Douglas before certification. A senior engineer approved the results for the company. Later, as a DER, 'the same engineer approved the report of the tests as acceptable documentation showing that the DC10 cargo door complied with the airworthiness regulations' (Eddy *et al.*, 1976: 180).

    Similarly, the pylon-engine design which was a contributing cause of the 1979 crash was approved by a DER, who reviewed his own design work as well as that of a colleague contained in a four-volume document. Following the accident, FAA investigators discovered that 'the second report was "incorrect" in part and differed from the first' (Green *et al.*, 1979:1)
14. Basnight explains: FAA engineers are 'bypassed in the sense of the state of the art, training and team effectiveness' (in Dean and

Kendall, 1979: 3). 'Although the FAA recruits and hires competent people, 'technology changes and that technology doesn't fit the FAA guy's background. He needs to be constantly upgraded or broadened in dimension or given new insight' (Dean and Kendall, 1979: 3).

15. A risk analysis

> is useless unless the exact nature of what is being predicted is known, reasonable, and relevant. By only retaining authority to approve the 'conclusions' of a probability value, FAA has abandoned . . . [its] capacity to enforce those FARs which contain an option to show compliance through extreme improbability . . . In order to assure compliance with the regulations, not only must the FAA be aware that risk analyses are being used but must also review the basic assumptions, logic, and methodology of those analyses. (United States House of Representatives: 1980c: 21–2)

16. Much of the following discussion comes from Eddy *et al.* (1976) and Johnston (1976).
17. Boeing was absorbed with its 747; Douglas was preoccupied with a stretched version of its DC8, 'straining to deal with DC8 and DC9 backlog', and in the middle of a company merger and personnel realignment (Eddy *et al.*, 1976: 71)
18. Eastern won a $24.5 million settlement against Douglas in this case (Eddy *et al.*, 1976: 72).
19. Electric actuators have fewer moving parts and are cheaper to maintain. Hydraulics provide an inherent safety feature though: if improperly secured and under pressure, a hydraulically-controlled door will 'ooze' open and allow a gentle loss of air, preventing the cabin from becoming fully pressurised. An electrically-powered door in the same situation is irreversible and will grip the fuselage tenaciously until the pressure differential causes a blow out.
20. L1011 and TriStar are interchangeable names for the Lockheed airbus aircraft.
21. Two reasons have been proffered for the choice: 1) General Electric built DC10 engines and offered to finance their purchase through Morgan Guaranty Bank, one of UAL's largest single shareholders. 2) UAL's concern that 'if the DC10 isn't built, then Douglas will be out of business . . . that would be a bad thing for the airlines, and a bad thing for the industry' (as quoted in Eddy *et al.*, 1976: 80).
22. There were design differences however. First and foremost they differed in design philosophies: Lockheed wanted to transfer its reputation for being on the leading edge of technology in its complex and novel defence aircraft designs to the commercial sector by incorporating as much new technology as possible in its wide-body, making it reliable, useful and maintainable in the process. Douglas took a more conservative approach to the DC10 design.

    The most visible difference is in the mounting of the Number Two, or tail, engine. The TriStar's is buried inside the rear of the fuselage and sucks its air in through a curved scoop which opens up just in front of the vertical tail fin. The Douglas engine is a part of the tail, well above the fuselage with a straight line access to air induction. Most aviation engineers feel the Lockheed-type tail engine is the safer of the two designs in terms of stability and handling qualities. However, it did take a great deal of time to design and develop a compatible engine.

Another area of design divergence is found at the rear of the aircraft where the pressurised section of the aircraft ends: the DC10 has a flat disk; the L1011 has a spherical segment curving toward the tail. The curved bulkhead requires greater time to design, but it is significantly lighter weight than the flat disk approach. Additionally, the TriStar walls are thinner and lighter than those of the DC10, due to the incorporation of new technology.

Finally, the flight deck windows differ. No matter how sophisticated the instrumentation, the pilot's visibility is always considered vital. 'See and avoid' is the phrase reflecting the pilot's responsibility to look out for other aircraft, or similar obstructions, in his area of flight.

> The established method – used in the DC10 and traceable back to the DC1 is to fit a number of flat glass plates into the upper nose, minimising as far as possible the inevitable discontinuities between flat surfaces and the compound curves of the streamlined skin. (Eddy *et al.*, 1976: 96)

Unfortunately, the necessary series of metal sheets interferes with the pilot's view. Insuperable optical problems are created when the windscreen is molded to airframe curves. A compromise is incorporated into Boeing 747s and Lockheed L1011s: conical or cylindrical sections formed to the windscreen basic shape. These are the product of time-consuming, new generation technology and are therefore more expensive to construct and replace. They provide the pilot with a much better view.

23. For those parts of the plane for which redundant systems cannot be designed, such as landing gear, the strategy shifts to safe-life. Thus landing gear would be built to last three times as long as normally expected.
24. Boeing first incorporated this level of redundancy in its 747s. The decision was rewarded in July 1971 when a Pan Am flight smashed into a pier at San Francisco's International Airport severing three of the four systems; the plane landed safely with only the fourth operating.
25. The size effect has to do with the dynamics of larger volumes of pressurised air in the jumbos. When airframe dimensions are doubled, air volumes are cubed, filling cabins with eight times as much air as before. The internal pressure differential rises dramatically, and requires a compensating design feature, such as greater floor strength or more venting. Aircraft are always built with venting systems in cabin floors to allow for pressurisation re-stabilisation should decompression occur as a result of a hole in the airframe. They also have pressure relief valves which prevent excessive pressures between the cabin and the outside. Cabin floors were designed to deal with a pressure differential of 3 pounds per square inch. (Subsequent to the 1974 crash, floors were

strengthened.) That being the case, the venting had to be adequate to allow for the greater differential caused by sudden decompression; cabin air needed to be able to escape nearly as fast as the air in the cargo hold.

26. The FAA established this standard using a simple logic; it amounted to the area between 2 frames and 2 longerons. The frames and longerons were the aluminium skeleton that gridded the plane. Skin was laid over the skeleton and riveted on, creating thousands of small rectangular areas, or panels, defined by crisscrossing members. If the fuselage were pierced, or if it cracked from fatigue, the skin might fly back in the rush of decompression, but the rivets would contain the ripping of the skin (Johnston, 1976: 225).
27. In an effort to reduce costs by pooling resources for maintenance, repair, spare parts stockpiling, and so on, European airlines purchasing Boeing 747s created the ATLAS consortium. Its members are Air France, Lufthansa, Alitalia and Sabena. Later, airlines purchasing McDonnell Douglas DC10s created the KSSU consortium with KLM, SAS, Swissair and UTA (French) as members. Many ATLAS members now also operate DC10s.
28. The simplest and most inherently safe door is the 'plug type', which opens inward and is larger than its frame. When closed and the cabin interior pressurised, this door is impossible to open because the pressure pushes it more firmly into place. However, they are heavy and do not become a part of the load carrying structure when closed, which requires massive reinforcement around the door frames and adds more weight. Also they open inward eliminating potential cargo space.

    The second door type is a 'tension latch' door used by Boeing since the KC135. It opens outward, is comparatively light weight, and when closed becomes a part of the load sharing structure of the aircraft. It utilises hydraulics, a series of C-latches and consequential locking devices that are fail-safe. It is both economical and safe.

    > Up to the point in 1974 when the Boeing 747 had completed two billion passenger hours, there had been seventeen recorded failures of the door latching mechanism. In each case, the 'fail-safe' elements in the system alerted the crew in time. (Eddy *et al.*, 1976: 128)

    The third door type, used in L1011s, incorporates an elaborate system known as 'semi plug' door. It opens outward and utilises electric actuators, but once closed, functions as a plug door in that pressure inside the fuselage acts to make the closure more secure. Its disadvantages are a little added weight and the time required for its design. Lockheed used 150 design engineers on door systems (Eddy *et al.*, 1976: 129).
29. Nine specific conditions were identified under which '"the door will open in flight resulting in sudden depressurisation and possible structural failure of the floor"' (as quoted in Johnston, 1976: 230).
30. The FMEA analysed a somewhat earlier version of the DC10 door and Douglas management was eager to keep the production process moving, so Convair's conclusions, though deadly

accurate, were dismissed as inapplicable to the present situation.

31. Even if the incident could be fairly blamed on human error, C.O. Miller, President of Safety System, Inc., and a consultant in accident prevention, explains that 'human error' is most often design-induced.

    It does not follow that, if an accident is ascribed to pilot error, controller error, or anybody else's error, the best remedial action rests with the person or persons in the same category. It may be the only short-term solution available; but a much more time- and cost-effective approach is to look for the design approaches that can be used to minimise if not eliminate the problem over the long run. (Johnston, 1976: 131)

32. A fully loaded plane would have added seven tons.
33. Boeing's 707 market success was attributed, in part, to its being first on the market. This had relegated Douglas to second place. Eastern's lawsuit for late deliveries of the DC9 had underscored the importance of on-time delivery.
34. Routine Service Bulletins are printed on white paper and are not related to airworthiness. Alert Service Bulletins are printed on blue paper and do concern safety problems, but not those affecting airworthiness.
35. The modifications which finally resulted in a fail-safe cargo door were detailed in four Service Bulletins issued by McDonnell Douglas. The first, 'SB52-27', was issued before Windsor recommending 'that the power supply to the electric actuators be rewired in a heavier gauge of wire . . .' Subsequent to Windsor, an additional recommendation was made to

    install a placard [in English] on each cargo door warning baggage handlers not to use more than fifty pounds of force when closing the handle: precisely how a baggage handler was supposed to measure his muscle was not explained. (Eddy *et al.*, 1976: 155)

    The second, 'SB52-35', was issued as an Alert Service Bulletin and called for the installation of a peephole and a decal showing what the baggage handler should see if the door were properly closed. It was questionable how well this would work because a)

    the door sill of a DC10 stands some fifteen feet above the ground, and each locking pin is less than two inches long. To make a proper inspection, each baggage handler would have to wait for the door to come down, and then move his mobile platform along to peep into the one-inch peephole. (Eddy *et al.*, 1976: 153)

    b) at night a flashlight would be needed, c) bad weather would discourage proper inspection because of the uncomfortable conditions and extra time required to do the job, and d) a window streaked with oil, dirt and water would make effective inspection dubious.

    Third, 'SB52-37', issued as routine, called for adjustments in the locking pin linkages and for the fitting of a support plate to 'make it impossible to close the door improperly'. The routine nature of the bulletin meant few DC10s were modified with alacrity. When the bulletin was issued 39 DC10s were in service: five were modified within 90 days; and one 'owned by National, was still flying around without a support plate on 5 March 1974 – nineteen months after the bulletin was issued' and two days after the Paris crash (Eddy *et al.*, 1976: 157).

Finally, 'SB52-49' was implemented to comply with the AD issued after Paris. This required the incorporation of a 'closed loop' locking system which made the locking system consequential. Now the cargo door was fail-safe.

36. The Committee concluded that

> While there is no longer any doubt that it is safe, it is an inelegant design worthy of Rube Goldberg . . . The DC10 intercompartment structures were not designed to cope with or prevent failures that would interfere with the continued safe flight and landing after the sudden release of pressure in any compartment due to the opening of a cargo door. It was therefore incumbent on McDonnell Douglas to show that loss of the cargo door . . . was 'extremely remote' for compliance with Federal Airworthiness Regulation 25 . . . It would appear, in light of the two accidents, that the level of protection and reliability provided in the cargo door latching, safety locking mechanisms, and the associated warning systems was insufficient to satisfy the requirements of FAR 25. It would appear, in light of the two accidents, that the level of protection and reliability provided in the cargo door latching, safety locking mechanisms, and the associated warning systems was insufficient to satisfy the requirements of FAR 25. Additionally, it now appears that the possibilities of improper door operations were not given adequate consideration for compliance with FAR 25. (as quoted in Eddy *et al.*, 1976: 160–1).

Regarding the use of a Gentleman's Agreement subsequent to Windsor, the report said:

> Review of the FAA's . . . correction programs associated with the DC10 airplane has again pointed out that the agency has been lax in taking appropriate Airworthiness Directive action where the need for ADs are clearly indicated. This situation is by no means unique to the DC10 airplane or to the Western Region. So-called 'voluntary compliance' programs have become commonplace on a large number of aeronautical products and in most, if not all, of the [FAA] regions. *The number of ADs issued is most likely to be considered a sales deterrent by the manufacturer.*
>
> Many complaints on this issue have been received from a number of foreign airworthiness authorities who have airworthiness responsibility over US manufactured products in their respective countries; and, for the most part, the FAA has ignored those complaints (as quoted in Eddy *et al.*, 1976: 160–1; emphasis is mine).

In particular reference to the Windsor incident, the report concluded

> The Agency was not effective in attaining adequate fleet-wide corrective action on a timely basis after problem areas were clearly indicated by the . . . accident. Non-regulatory procedures and agreements were used in lieu of established regulatory AD procedure [and] in the long run proved to be ineffective in correcting design defficiencies . . . to prevent reoccurrence of the accident. (as quoted in Eddy *et al.*, 1976: 160–1)

37. The grounding caught the airlines by surprise. Western actually thought it was a hoax at first. Planes and passengers were stranded all over the world. US operators included National Airlines, Northwest Airlines, World Airlines, American Airlines, Western Airlines, United Airlines and Continental Airlines. Overseas, the aircraft was flown by Alitalia, Iberia, Japan Airlines, KLM Royal Dutch Airlines, Korean Airlines, Laker Airways, Lufthansa, Sabena Belgian World Airlines, Scandinavian Airlines System, Swissair, Thai Airways, Turkish Airlines and Varig Brazilian Airlines.
38. Because the DC10 is a heavier aircraft and carries a larger payload, it generates more revenues. Therefore, senior pilots generally chose to fly these aircraft and earn more money, not only by virtue of their seniority, but also because of the profit the aircraft generates. To be able to fly other planes, pilots must participate in an expensive two-month retraining period. No one in the industry expected the grounding to last more than two weeks, so airline management was reluctant to make the decision to pursue this tactic.
39. These are also areas where the DC10 differs from its rival L1011 airbus.
40. In the TriStar and 747, wing control hydraulics lines are routed along the trailing (back) edge, where they are less vulnerable to damage. The L1011 and B747 have mechanical braking systems to prevent sudden slat retraction. The L1011 has alternate, redundant stall warning and slat disagreement devices.
41. The monoball failures were caused by poor quality control production procedures which had resulted in use of substandard materials.
42. Originally, this saved 30 labour hours per plane. Once personnel became efficient the savings rose to 200 hours per plane.
43. Once at the Los Angeles International Airport and once at the American's new maintenance headquarters in Tulsa, Oklahoma.
44. 19 December 1978 and 22 February 1979.
45 This raises an interesting question: Would more strict initial surveillance and preventive measures after Chicago's accident have kept the aircraft flying, thereby imperilling the public, or would it have resulted in a more immediate and equally as safe resolution to the problem as what followed the grounding anyway?
46. The Association of European Airlines (AEA) was established in Munich on 14 September 1973, to facilitate co-operation among European carriers. Nineteen major European scheduled airlines are members: Air France, Aer Lingus, Alitalia, Austrian

Airlines, British Airlines, British Caledonian Airways, Finnair, Iberia, Icelandair, Yugoslav Airlines, KLM, Lufthansa, Luxair, Olympic Airways, Sabena, Scandinavian Airlines System, Swissair, Turkish Airlines, TAP-Air Portugal and UTA. Its headquarters are in Brussels.

According to its statutes, the AEA is required to 'contribute to the improvement and development of European commercial air transport in the interest of the public and the member airlines' (draft copy of 'Origin of AEA and What It Does', enclosed in Hayward, 1982: personal correspondence). It participates in the Joint Steering Committee of the European Joint Airworthiness Requirements, co-operates with other industry groups such as the Air Transport Association, and acts as a liaison between member airlines and other multinational organisations such as ECAC and IATA.

The AEA's Technical Affairs Committee (TAC) has study groups which often collaborate with corresponding working groups of ECAC's technical committee in order to issue joint recommendations to Member States. The TAC was directly involved in the early recertification of European DC10s following the 1979 DC10 crash.

The European Civil Aviation Conference (ECAC) was created in 1955 to co-ordinate intra-European air transport. Its aims and objectives are to review the development of European air transport in order to promote the co-ordination, better utilisation and orderly growth of Europe's airways, and to consider any special problem which might arise. There are 22 Member States, of which 19 held original membership: Austria, Belgium Cyprus, Denmark, Finland, France, Germany (Fed.Rep.), Greece, Iceland, Ireland, Italy, Luxembourg, Malta, Netherlands, Norway, Portugal, Spain, Sweden, Switzerland, Turkey, United Kingdom and Yugoslavia. It was also involved in the DC10 recertification.

47. Reflecting the important role played by politics in the whole affair, the US delegation was chaired by James R. Atwood, Deputy Assistant Secretary of the State Department; other members included FAA Administrator Langhorne Bond and experts from the State Department, FAA, Department of Transportation and the Civil Aeronautics Board. The ECAC delegation was headed by Erik Willock, Director-General of Civil Aviation of Norway, acting in his capacity as President of ECAC; he was assisted by Kaarlo J. Temmes, Director-General of Civil Aviation of Finland and Chairman of the ECAC Technical Committee, and experts from nine ECAC Member States – Belgium, Denmark, Finland, France, Netherlands, Norway, Spain, Turkey and the UK.
48. Indonesia's Garuda (19/6/79), Air New Zealand (22/6/79), Pakistan International Airlines (25/6/79), Malaysian Airline Systems (4/7/79).
49. European DC10s were longer-range Series 30 and Series 40 aircraft; US airlines flew Series 10 aircraft.
50. In personal correspondence, Chaplin commented that 'the existence of the JARs system . . . enabled us more easily to work as a team in 1979'

(1983); Bouma, that 'an important side-effect' of 'the cooperation in the JAR system . . . was the unified position most European Airworthiness Authorities took relative to the suspension by FAA of the DC10 Type Certificate and subsequent airworthiness measures' (1983); Kennedy, that 'one of the "spin-offs" of the JAR activities is that the authorities in Europe now meet frequently and know each other much better than was the case and collaborated on the DC10' (1983).

5 Conclusions: Structure, Process and Reciprocity

1. We need a reciprocal comprehension of 'how domestic politics affect patterns of interdependence and regime formation' and 'how economic interdependence and institutions such as international regimes affect domestic politics' (Keohane and Nye, 1978: 748). David Lake observes 'structural theory . . . lacks a conception of process, or an explanation of how the constraints or interests derived from the international economic structure are transformed into decisions or political strategies within particular countries' (1983: 539). Peter Katzenstein would add that we need to understand how selected policies or interests derived from within particular countries are transformed into new constraints and interests at the global level (1976). Christer Jönsson suggests we add a 'process model' to existing 'structural' and 'functional' models, so we can link our knowledge of changes in international structures or issue area dynamics to our frustration at dealing 'with collective irrationality resulting from individual rationality' by being aware of what was available for negotiation and what was compromised (1987b: 26; see also 1986).
2. Here I am using the following definitions: (a) 'A system is a set of interacting units having behavioral regularities and identity over time' (Nye, 1988: 241); (b) structure is used to convey a sense of 'the unplanned and unchanging nature of international political and economic realms' (Strange in Ikenberry, 1986: 65,fn.25), keeping in mind the admonition that even structures, like 'coral reefs', do change (Wallerstein, 1976); (c) the term 'nonstructural' is also borrowed from Nye and used here to connote variables which are less rigid or permanent and more easily subject to change; (d) finally, the 'international system is understood to be an "anarchic society" only in the sense that there is no formal government above states. There exists, however, communication, cooperation and governance, the latter generally embodied in a series of regimes managing issue areas' (see Nye, 1988: 249; Bull, 1977).
3. See Jonathan Aronson (1977), Charles Lindblom (1977) and Terry Moe (1980), among others.
4. Jonathan Aronson distinguishes between direct and indirect power. In contrast to direct power where actor A induces actor B to do something it would not otherwise have done (Dahl, 1963: 25–6, 47–8), indirect and circumventing power are subsumed under a

type of influence which 'entails the ability of an actor to mold the decision environment of another actor, thereby limiting the scope of options available to the second actor . . . ' even though the effects may be unintentional. More specifically, circumventing power 'allows actors to use their international structural flexibility and legal loopholes . . . ' to evade regulations (or their effects) ' . . . but may or may not be designed to influence government policies and choices' (1977: 18–19).

5. See Keohane, 1984: 97; Keohane and Nye, 1977: 55–7; Krasner, 1982b: 499; Lipson, 1982: 453; Stein, 1982: 322–3; Young, 1982: 280.
6. James Rosenau talks about two kinds of learning leading to change:

   > (a) when the external stimuli are so persistently and startlingly different as to jolt habitual modes and foster new patterns more appropriate to the evolving circumstances; or (b) when new skills, capabilities and/or responsibilities develop within the actor, forcing the old, habitual ways to yield to new ones (1986: 864).

   This is similar to my argument that the latter comes first, making the former eventually possible if not inevitable.
7. See Keohane and Nye, 1987: 749–52 for a discussion of this aspect of 'learning'.
8. In the words of Peter F. Cowhey and Edward Long, 'great powers institute regimes to their liking that will not prove totally unacceptable to key partners' (1983: 160,fn.5).
9. Standards are a double-edged sword: on the one hand, they can facilitate trade and economic growth by decreasing inefficient costs of duplication (Kindleberger, 1978; Smith, 1776); on the other hand, they facilitate market manipulation as a means of coercion or rule enforcement (see McKeown, 1983: 77–8; 1986).
10. Restated, Russett's assertion regarding state elites fits corporate elites as well: corporate elites 'experience incentives to incur long-term costs [payable after they have left the corporation protected by a golden parachute] to show short-term gains, or at least to avoid short-term losses'. More than one corporate executive has commented to me that they do not like the loss of control they associate with the proliferation of transnational joint venture projects. Nevertheless their choices are few given technology, industry dynamics and market structure. And one thing is certain, without a short-term there is no long-term.
11. A number of historical case studies supports this contention. See for example, Cohen, 1977; Gilpin, 1975, 1981; Kindleberger, 1973; Keohane and Nye, 1974, 1977; and Krasner, 1976.
12. For example, natural rubber was replaced by synthetic materials, decreasing the value of rubber plantations as power resources. The intrinsic value of oil, on the other hand, increased as Western powers became more dependent on it to fuel economic growth. State elites, in whose territory the crude oil lay, also learned to use oil more effectively as a bargaining chip. Finally, technological innovation and application took on increased significance as a new resource.
13. Victoria Curzon-Price has suggested that, in trying to predict when

change might occur, general economic conditions must be considered. When there is a sense of progress or purpose among global actors or when there is domestic prosperity within dominant countries, there is a greater willingness to respond to international constraints and injunctions. However, when there is stagnation or a sense of relative decline, politicians respond to constituent demands for change (in Strange and Tooze, 1981).

14. This perception was so dominant that even the Reagan administration saw this as a prime target for spending cuts in its early efforts to balance the budget.
15. The DC10 was not a product of purely US domestic input, but nothing as expensive or visible as an engine was imported from a foreign source.
16. With their own regulations, any costly delays in certification would be the result of their own standards and procedural requirements. Difficulties surrounding the Concorde's certification in the United States emphasised the need for European co-operation in this arena to avoid manipulation. Although Concorde may have been a market-place failure anyway, its success was doomed by US aircraft manufacturers and environmentalists who utilised US regulations to prevent the European supersonic transport (SST) from exercising its primary advantage over other aircraft – speed – within US territory. The motivations differed, US manufacturers were not allowed to build their own SST and environmentalists wanted to protect Americans' quality of life. The result was US market manipulation. European aircraft manufacturers were not in control of their own destiny.
17. Perhaps one of the most dramatic examples of this is found in the transition from building the Concorde to construction of the Airbus.
18. Behaviour within the manufacturing segment of commercial aviation might indicate a declining hegemon because of the proliferation of transnational, joint production efforts in an industry once jealously guarded for its usefulness to national security.

    Behaviour within the service sector in the 1980s tells a different short-term story. It now appears that the United States will be able to impose its preference for an internationally economically deregulated air transport industry on the global community. In response to this strategy domestically, airlines rationalised production, creating a few strong airlines better able to compete with other airlines internationally. These other airlines now feel compelled to follow suit. This is a curious trend when we remember Krasner's observation that states could exercise more control over the industry within an authoritatively allocated regime. That the United States can exercise such power is impressive; that it is a desirable long-term strategy for the United States in its attempt to maintain dominance in the regime is questionable. This may result in another failed attempt at veiled protectionism. As global airlines rationalise production (once thought impossible but now conceivable, see Jönsson, 1981), these airlines may offer stronger competition than US government elites anticipated. Again, the European Community has learned that it is possible to generate and reap benefits from transnational ventures. As Community members move steadily towards 1992, such co-operation seems more likely.

19. See for example, Claude, 1964; Groom and Taylor, 1975; Haas, 1964; Mitrany, 1943, 1948; Taylor, 1968.

Bibliography

Aviation Safety

'ACAP Asks Congress to Investigate Head of FAA' (1983) *ACAP: A Quarterly Report* (December): 5, Washington, DC: Aviation Consumer Action Project.

'ACAP Charges FAA Fails to Act in Two Years Since DC10 Crash' (1981) *News from ACAP* (June): 1–2. Washington, DC: Aviation Consumer Action Project.

'Airline Pilots Are Alarmed Over How the Airworthiness Certification Process is Administered: Why?' (1980) Washington, DC: Air Line Pilots Association.

'Airline Supports Recent APA Petition on DC10' (1980) *APACE: Airline Passengers Association Newsletter* 4, 5 (October–November): 1.

Airworthiness Committee (1980) Report of Thirteenth Meeting, 29 October to 9 November; DOC 9293, AIR C/13. Montreal: International Civil Aviation Organisation.

ALLWARD, M. (1967) *Safety in the Air* New York: Abelard-Schuman, Ltd.

'ALPA Criticizes FAA Conduct of Aircraft Certification Role' (1978) (Press Release 78.23, 15 May) Washington, DC: Air Line Pilots Association.

ASHFORD, R. (1986) 'Aviation Safety – The Economic Impact of Airworthiness Regulation'. Paper presented at the Flight Safety Foundation 39th International Air Safety Seminar (6–9 October).

The Aviation Safety Institute (1979) Worthington, OH: Aviation Safety Institute.

BARCLAY, S. (1970) *The Search for Air Safety* New York: William Morrow & Company, Inc.

BEARD, M.C. (1982) *Export/Import Airworthiness Certification of Civil Aeronautical Products* (1 March) Washington, DC: FAA Office of Airworthiness.

BEBCHICH, L.N. (1980) 'Brief for Petitioner in No. 79–1662, British Caledonian Airways, Limited' United States District Court of Appeals for the District of Columbia Circuit. Washington, DC: Martin, Whitfield, Smith, and Bebchich.

BECK, M., W.J. COOK, S. DOHERTY, M. HAGER, H. MORRIS, M. RESENER, and R.SANDZA (1984) 'Can We Keep the Skies Safe?' *Newsweek* 103, 5 (20 January): 24–31.

'Boeing, FAA Seek Answer to Observer Seat Problem' (1983) *Aviation Week and Space Technology* 118, 9 (28 February): 29.

CARLEY, W.M. (1979a) 'Danger Aloft: Despite Major Gains, Airlines Still Present Mechanical Hazards' *Wall Street Journal* (1 November): 1.

'Changes Proposed to Avert Slat Asymmetry on DC10' (1982) *Aviation Week and Space Technology* 116, 2 (11 January): 30–2.

'Check of DC10s Ordered for Fire Hazard' (1984) *Denver Post* (May 26): 17A.

'Civil Aircraft Accident Investigation Guidelines' (1980) Washington, DC: National Transportation Safety Board.

The Continuing Airworthiness of Aircraft in Service – Methods of Handling and Exchange of Information on Airworthiness Directives (or Their Equivalent) (1981) Circular 95-AN/78/4 (Fourth Edition, September) Montreal: International Civil Aviation Organisation, *Convention on International Civil Aviation*

'Council Directive of 16 December 1980 on Further Cooperation and Mutual Assistance Between the Member States in the Field of Air Accident Investigation' (80/1266/EEC). *Official Journal of the European Communities* L375, 23 (31 December 1980): 32–3.

'Crew Decisions Cited in Crash of 737' (1982) *Aviation Week and Space Technology* 117, 13 (27 September): 78–89.

'Danger: Fire in the Cargo Compartment' (1983) *News from ACAP: A Quarterly Report* (December): 4.

EPSTEIN, A. (1984) 'FAA Protected from Lawsuits Over Inspection', *Denver Post* (20 June): 15A.

EUROPEAN COMMUNITIES (1987) 'Report drawn up on behalf of the Committee on Transport on Community measures in the field of air transport safety' European Parliament Session Documents, Series A, Documents A 2-135/87 and A 2-135/87/B (3 August).

European Parliament Working Documents 1981–1982: Motion for a Resolution . . . on Safety Measures in Aircraft (Document 1-701/81 PE 75.875) (6 November 1981) Brussels, Belgium: European Communities.

'FAA Aircraft Certification is "Closed" Despite Counter Pledge of Open Government' (1978) Air Line Pilots Association Press Release 78.48 (26 September) Washington, DC: ALPA.

'FAA Issues Proposals to Increase and Decrease Safety' (1981) *News From ACAP* (June): 3–4.

'FAA Reorganises Flight Standards Unit' (1979) *Aviation Week and Space Technology* 111, 16 (16 August): 32–3.

'Factors in Air Florida Crash Cited' (1982) *Aviation Week and Space Technology* 117, 14 (4 October): 65–75.

FEAVER, D.B. (1984) 'Jetliner Flight Surge Sets Off Safety Fears' *Denver Post* (21 June): 1A–3A.

FINUCANE, M. (1983) 'Deregulation of Air Safety Under Reagan: Flying in Fear' *News from ACAP: A Quarterly Report* (December): 1, 6–7.

FROMM, G. (1969) *Aviation Safety* Washington, DC: The Brookings Institute.

'The Future of Aviation: Report of the Fourth International Workshop on the Future of Aviation' (1986) *Transportation Research Circular* No. 299 Transportation Research Board (8–10 October 1985) Washington, DC: National Academy of Sciences.

'The Future of Aviation: Report of the Fifth International Workshop on the Future of Aviation' (1988) *Transportation Research Circular* No. 329 Transportation Research Board (6–8 October 1987) Washington, DC: National Academy of Sciences.

GALIPAULT, J.B. and I.R. RIMSON (1979) 'Incident Analysis as a Key to Accident Prevention' *Forum* (Winter), Washington, DC: International Society of Air Safety Investigators.

'GAO Hits FAA Safety Capabilities' (1980) *Aviation Week and Space Technology* 112, 10 (3 March): 34.

GLINES, C.U. and M. MORRIS (eds) (1979) *No Compromise with Safety: The Crew Complement Question* Paris, France: Europilote and Washington, DC: ALPA.

GOLICH, V.L. (1986) 'An Assessment of Deregulation's Impact on the US Airline Industry', Unpublished report prepared for the Colorado Public Utilities Commission (14 November).

—— (1988) 'Airline Deregulation: Economic Boom or Safety Bust?' *Transportation Quarterly* 42, 2 (April): 159–79.

HAY, G.C., C.D. HOUSE, and R.L. SULZER (1978) *Summary Report of 1977-1978 Task Force on Crew Workload* United States Department of Transportation, Federal Aviation Administration, Office of Systems Engineering Management, Report No. FAA-EM-78-15, Springfield, VA: National Technical Information Service.

HELMS, J. (1981) *Semiannual Report to Congress on the Effectiveness of the Civil Aviation Security Program*, 1 July – 31 December 1980 (15 April) Washington, DC: Department of Transportation, FAA Office of Aviation Policy.

'IATA's Technical Activities' (1982) *IATA Backgrounder* Montreal, Canada: International Air Transport Association, Public Information Department.

International Federation of Airworthiness Membership List (1982) England: International Federation of Airworthiness.

'An International Organization for Everyone Concerned with Improving the Safety of Flight' (1981) Arlington, VA: Flight Safety Foundation, Inc.

International Standards – Airworthiness of Aircraft: Annex 8 to the Convention on International Civil Aviation (1973) 8/73, E/P1/6000; 8/79, E/P2/1500, Annex 8/6 (Sixth Edition, July), Montreal: International Civil Aviation Organisation.

International Standards and Recommended Practices – Operation of Aircraft: Annex 6 to the Convention on International Civil Aviation, Part I – International Commercial Air Transport (1973) AA/72, E/P1,6500; 4/77, E/P2/3000; 2/81, E/P3/1500, Annex 6, Part 1 (Third Edition of Part I, October) Montreal: International Civil Aviation Organisation.

JACKSON, R.L. (1984) 'Panel Says Crew Misjudged Intense Jet Fire that Killed 23' *Denver Post* (11 July): 17A.

KENNEDY, A.P. (1979) 'Joint Airworthiness Requirements, Their History and Progress' *Aircraft Engineering* (May).

KING, J.B. (1978) 'Title 49 – Transportation; Chapter VIII – National Transportation Safety Board; Part 821 – Rules of Practice in Air Safety Proceedings' *Federal Register* 42, 250 (28 December): FR Doc. 78-36117.

—— (1979) 'Title 49 – Transportation; Chapter VIII – National Transportation Safety Board; Part 831 – Aircraft Accident/Incident Investigation Procedures' *Federal Register* 44, 116 (14 June): FR Doc. 79-18577.

—— (1980) 'Title 49 – Transportation; Chapter VIII – National Transportation Safety Board; Part 830 – Notification and Reporting of Aircraft Accidents or Incidents and Overdue Aircraft, and Preservation of Aircraft Wreckage, Mail, Cargo, and Records', *Federal Register* 45, 178 (11 September): FR Doc. 80-27741.

LEDERER, J. (1939) *Safety in the Operation of Air Transport* Northfield, VT: Norwich University.

—— (1982) 'Aviation Safety Perspectives: Hindsight, Insight, Foresight' Nineteenth Wings Club 'Sight' Lecture, Presented at the Wings Club on 21 April 1982, New York City.

LENOROVITZ, J.M. (1979) 'ALPA Plans Certification Role' *Aviation Week and Space Technology* 111, 21 (19 November): 25–6.

LOW, G.M., *et al.* (1980) *Improving Aircraft Safety: FAA Certification of Commercial Passenger Aircraft*, Washington, DC: National Academy of Sciences.

LOWELL, J.W. (1967) *Airline Safety is a Myth*, New York: Bartholomew House.

McLUCAS, J.L. (Chair) (1981) *Report of the President's Task Force on Aircraft Crew Complement* (2 July) Washington, DC.

MILLER, C.O. (1976) 'The Design-Induced Part of the Human Error Problems in Aviation', *Journal of Air Law and Commerce* 42, 1: 119–31.

Multilateral Agreement Relating to Certificates of Airworthiness for Imported Aircraft (1960) ICAO DOC. 8056 (19 April) Montreal: International Civil Aviation Organisation.

NADER, R. (1979a) Statement to the Subcommittee on Government Activities and Transportation of the House Committee on Government Operations (18 June) Washington, DC.

—— (1979b) Letter to Secretary of Transportation, Brock Adams, 27 May.

NATIONAL TRANSPORTATION SAFETY BOARD (1980) 'Title 49 – Transportation; Chapter VIII — National Transportation Safety Board; Part 800 – Organisation and Functions of the Board and Delegations of Authority', *Federal Register* 42 (1 March): FR 8379.

—— (1981) *Annual Review of Aircraft Accident Data: US Air Carrier Operations – 1979* (NTSB-ARC-81-1) (16 November) Washington, DC: Bureau of Technology.

'NTSB Focuses on Deicing Procedures' (1982) *Aviation Week and Space Technology* 117, 17 (25 October): 74–83.

'NTSB Investigates Runway Conditions' (1982) *Aviation Week and Space Technology* 117, 8 (1 November): 92–5.

'NTSB Reports Raises 737 Crash Issues' (1982) *Aviation Week and Space Technology* 117, 6 (18 October): 73–9.

NYSMITH, C.R. (1980) *Aviation Human Factors: The NASA Perspective*, paper presented at the 1980 FAA Human Factors Workshop, Boston, MA.

O'BRIEN, John (1987) 'Deregulation and Safety: An Airline Pilot's Perspective' (June) paper presented at Northwestern University Transportation Safety Conference, Chicago, IL.

O'DONNELL, J.J (1978) 'Aircraft Certification: Changes Must be Made'*Air Line Pilot* (November): 6–7.

—— (1979) 'Procrastination + Lethargy = Tragedy' *Air Line Pilot* (December): 10–2.

OTT, J. (1982c) 'House Units Scrutinize Safety' *Aviation Week and Space Technology* 117, 7 (16 August): 27–9.

—— (1988) 'Military Avoids US Carriers That Fail Safety Standards' *Aviation Week and Space Technology* (8 February): 98–101.

PEAR, R. (1981) 'Pilot Union Chief Calls Flying Safe in Spite of Strike', *New York Times* (20 August).

PERROW, C. (1984) *Normal Accidents: Living with High Risk Technologies*, New York: Basic Books.

'Pilots Union Endorses Safety of Air Travel' (1981) *Aviation Daily* (20 August).

'Pilots Union Chief Calls Airways "Safe," Says Near-Misses Are Below A Year Ago' (1981) *Wall Street Journal* (20 August).

POWER-WATERS, Capt. B. (1972) *Safety Last – The Dangers of Commercial Aviation: An Indictment by an Airline Pilot*, New York: The Dial Press.

Recommendations on Flight Safety (1971) INTS/3-1, 2, 3, 4, Paris, France: European Civil Aviation Conference.

Report by the Comptroller General of the United States: How To Improve the Federal Aviation Administration's Ability to Deal With Safety Hazards (1980) DOC CED-80-66 (29 February 1980) Washington, DC: General Accounting Office.

'Safety Board Issues Recommendations' (1982) *Aviation Week and Space Technology* 117, 21 (22 November): 85–91.

Safety in Flight Operations: Volume 1 (1975) Summary of Discussions 20th Technical Conference, Istanbul, Turkey, 10–14 November, Montreal, Canada: International Air Transport Association.

'Suit Seeks Airline Carriage of Emergency Medical Gear' (1982) *Aviation Week and Space Technology* 117, 22 (29 November): 41.

UNITED STATES DEPARTMENT OF TRANSPORTATION (1981) *Airworthiness Overview Manual* (Draft, April) Washington, DC: Federal Aviation Administration Office of Airworthiness.

—— (1984) *National Air Transportation Inspection Program* (July) Washington, DC: Federal Aviation Administration.

UNITED STATES HOUSE OF REPRESENTATIVES (1975) *Transportation Safety Act of 1974*, Washington, DC: US Government Printing Office.

—— (1978) *FAA Certification Process and Regulation of Illegal Commercial Operators, Volume I* Hearings before a Subcommittee of the Committee on Government Operations; 95th Congress, 2nd Session (26–28 September) Washington, DC: US Government Printing Office.

—— (1980a) *FAA Aviation Safety Issues, Volume II.* Hearings before a Subcommittee of the Committee on Government Operations, 96th Congress, 1st Session (13–14 August 1979) Washington,DC: US Government Printing Office.

'What Congress Says About the FAA's Certification Procedures' (1980) Washington, DC: Air Line Pilots Association.

Case Studies

Accident Report – American Airlines, Inc. (1979) DC10-10, N110AA, Chicago O'Hare International Airport, Chicago, Illinois, 25 May 1979 (NTSB-AAR-79-17) Washington, DC: National Transportation Safety Board.

'Airlines and Aviation Officials Continue Efforts to Cope With Grounding of DC10s' (1979) *Wall Street Journal* (18 June): 29.

Annual Report to Congress – 1979 (1979) Washington, DC: National Transportation Safety Board.

CARLEY, W.M. (1979b) 'Behind the Crash: FAA Inquiry Changes DC10 Manufacturing, Maintenance Mistakes', *Wall Street Journal* (17 July): 1, 25.

'DC10 Inspections Turn Up 37 Planes With Safety Flaws' (1979) *Wall Street Journal* (1 June): 26.

DEAN, P. and J. KENDALL (1979) 'Monitoring of Maintenance Data Criticized: DC10 Crash Could Have Been Averted, Expert Says' *Los Angeles Times* (23 June): 3.

EDDY, P., E. POTTER and B. PAGE (1976) *Destination Disaster – From the Tri-Motor to the DC10: The Risk of Flying*, New York: The New York Times Book Company.

'European Air Carriers Propose Joint Program on DC10 Maintenance' (1979) *Wall Street Journal* (13 June): 27.

EUROPEAN CIVIL AVIATION CONFERENCE (1979a) Press Release No. 43E (25 June) Paris, France: European Civil Aviation Conference.

—— (1979b) Press Release No. 41E (12 -June) Paris, France: European Civil Aviation Conference.

'European DC10's Return to Skies After Tough Maintenance Checks' (1979) *Los Angeles Times* (20 June): 10.

'Europeans Clear Inspection Program to Allow DC10 Use' (1979) *Wall Street Journal* (19 June): 25.

'FAA Grounds 20 DC10s, Saying Checks on Jets May Have Caused New Damage' (1979) *Wall Street Journal* (5 June): 3.

'FAA Issues Airworthiness Directive on Redundancy In DC10 Stall Warning' (1980) *Aviation Week and Space Technology* 112, 3 (21 January): 24–5.

'FAA Orders All DC10s Grounded For Safety Check' (1979) *Wall Street Journal* (30 May): 2.

'FAA Orders US Lines to Ground DC10s To Inspect Their Engine-Mounting Bolts' (1979) *Wall Street Journal* (29 May): 3.

'FAA Proposes DC10 Pylon Changes' (1980) *Aviation Week and Space Technology* 112, 4 (28 January): 26.

'FAA Suspends Design Certification, Grounds DC10s, Prompting a Dispute Over Cause of Plane's Problems' (1979) *Wall Street Journal* (7 June): 2.

'FAA Team Says Design of DC10 Pylon Makes a Service Hard, Posing Safety Risks' (1979) *Wall Street Journal* (11 July): 10.

FEAVER, D.B. and F. BARBASH (1979) 'Safety Systems on Other Jetliners Lacking on DC10' *Los Angeles Times* (24 June): 5.

'Foreign Airlines Resume Flying DC10s After a New Inspection Code Is Approved' (1979) *Wall Street Journal* (20 June): 5.

GIRARD, P. (1979a) 'Maker Defends DC10's Basic Systems' *Los Angeles Times* (4 August): 4.

—— (1979b) 'Crash Probers: Safety Board: Small Staff, Big Burden' *Los Angeles Times* (31 July): 1+.

GODSON, J. (1983) *Government and British Civil Aerospace: A Case Study in Post-War Technology Policy*, Manchester University Press.

GORE, R.J. (1979) 'Each DC10 Begins as a Pile of 270,000 Parts', *Los Angeles Times*, Part I (13 July): 1+.

GRANT, L. (1979) 'DC10 Grounding will Hurt Western, Continental – But How Much?' *Los Angeles Times* (19 June): I–IV.

GREEN, L., R.L. JACKSON, P. GIRARD, G. DOYLE, J. KENDALL, W.C. REMPEL, P. DEAN, and G. SHAW (1979) 'Signs Unnoticed: DC10 Tragedy: Flaws, Loopholes', *Los Angeles Times* (13 July): 1+.

HARRIS, R.J., Jr. (1979) 'Winging It: Continental Airlines, Its DC10s Grounded, Does a Juggling Act', *Wall Street Journal* (15 June): 1+.

JOHNSTON, M. (1976) *The Last Nine Minutes; The Story of Flight 981*, New York: William Morrow and Company, Inc.

KARR, A.R. (1979a) 'Danger Aloft: FAA Is Often Accused of Laxity and Delays in Its Air-Safety Role', *Wall Street Journal* (12 November): 1+.

KENDALL, J. (1979a) 'Memo Alleges Insufficient Data in DC10 Approval', *Los Angeles Times* (12 July): 11.

—— (1979b) 'Asymmetrical Slats Called Main Issue in Lifting DC10 Groundings', *Los Angeles Times* (3 July): 5.

KRONHOLZ, J. (1979) 'Freddie Laker Is Hit by DC10 Ban Too, But Looks to Return', *Wall Street Journal* (15 June): 13.

NATIONAL TRANSPORTATION SAFETY BOARD (1979) Letter to FAA Administrator Langhorne Bond, Safety Recommendation A-79-31, 9 May.

REMPEL, W.C. and L. GREEN (1979) 'DC10 Pylons Had Flaws Early On, Documents Show', *Los Angeles Times* (20 June): 1+.

RUFF, C.F.C. (1981) 'Brief for Respondent in Nos. 79-1662 and 79-1737. Langhorne M. Bond, Administrator, Federal Aviation Administration' DOJ-1981-02, United States District Court of Appeals for the District of Columbia Circuit, Washington, DC: Department of Justice.

SMITH, L. (1979) 'They've Turned Off the Seat Belt Sign at the McDonnell Douglas', *Fortune* (17 December): 60–4.

'Some Lessons from the O'Hare Disaster' (1979) *News from ACAP* (September): 1-2.

'Statement of Ralph Nader' (1979) Press Release, Aviation Consumer Action Project (25 May) Washington, DC.

'Swiss, Dutch Airlines to Resume DC10 Flights' (1979) *Los Angeles Times* (19 June): 7.

'United States and ECAC Representatives Meet for the Second Time on the DC10 Grounding Issue' (1979) Press Release No. 44E (4 July) Paris, France: European Civil Aviation Conference.

UNITED STATES DEPARTMENT OF TRANSPORTATION (1980a) *DC10 Decision Basis* Washington, DC: Federal Aviation Administration.

UNITED STATES HOUSE OF REPRESENTATIVES (1980b) *FAA Certification Process III*, Hearings before a Subcommittee of the Committee on Government Operations; 96th Congress, 1st Session (11,18 June; 9, 10 October 1979) Washington, DC: US Government Printing Office.

—— (1980c) *A Thorough Critique of Certification of Transport Category Aircraft by the Federal Aviation Administration*, Sixteenth Report by the Committee on Government Operations (7 May) Washington, DC: US Government Printing Office.

'What Have We Learned?' (1980) *APACE: Airline Passengers Association Newsletter* 4, 5 (October–November): 1.

History

Air Services Agreement Between the Government of the United States of America and the Government of the United Kingdom of Great Britain and Northern Ireland (1977) T.I.A.S., No. 8641 (23 July).

Air Service Agreement Between the Government of the United States of America and the Government of the United Kingdom of Great Britain and Northern Ireland (1946) 60 Stat. 1499, T.I.A.S., No. 1507 (11 February).

'Background to a Bilateral Row' (1971) *Flight* 100, 3273 (2 December): 887.

BOGOSIAN, R.W. (1981) 'Aviation Negotiations and US Model Agreement', *Journal of Air Law and Commerce* 46, 4: 1007–37.

BURDEN, W.A.M. (1945) *'Opening the Sky' Blueprint for the World Civil Aviation* (Publication 2348, Conference Series 70) Washington, DC: Government Printing Office.

Civil Air Policy: A Report by the Air Coordinating Committee (1954) Washington, DC: Government Printing Office.

COOK, D. (1945) *The Chicago Aviation Agreements: An Approach to World Policy*, New York: American Enterprise Association, Inc.

COOPER, J.C. (1947) *The Right to Fly,* New York: Henry Holt.

—— (1952) 'Air Navigation Conference, Paris 1910', *Journal of Air Law and Commerce* 19 (Spring): 127–43.

DAVIES, R.E.G. (1964) *A History of the World's Airlines*, London: Oxford University Press.

50 Years of World Airline Cooperation (1969) Montreal, Canada: International Air Transport Association.

FITZGERALD, G.F. (1978) 'International Civil Aviation Organization and Development of Conventions on International Air Law 1947–1978', *Annals of Air and Space Law* 3: 51–120.

GRIFFITHS, D.R. (1978b) 'Bermuda 2 Principles Dropped', *Aviation Week and Space Technology* 108, 4 (23 January): 31–2.

'The Inside Story of the Aviation Act' (1958) *American Aviation* (16 June).

KANE, R.M. and A.D. VOSE (eds) (1969) *Air Transportation*, Dubuque: William C. Brown Book Co.

LOWENFELD, A.F. (1975) 'A New Takeoff for International Air Transport', *Foreign Affairs* 54 (October): 36–50.

McCLURKIN, R.J.G. (1948) 'The Geneva Commission on a Multilateral Air Transport Agreement', *Journal of Air Law and Commerce* 15, 1: 39–46.

NAYLER, J.L. and OWER, E. (1965) *Aviation: Its Technical Development*, London: Peter Owen/Vision Press.

PILLAI, K.G.J. (1969) *The Air Net: The Case Against the World Aviation Cartel*, New York: Grossman Publishers.

'The President Transmits US-UK Financial Agreement to Congress' (1946) *The Department of State Bulletin* 14, 345 (10 February): 183–4, 216, Washington, DC: US Government Printing Office.

Public Law 95–504, 92 Stat. 1706, 49 USC. 1302, (Airline Deregulation Act of 1978).

Public Law 89-670, 80 Stat. 931, 49 USC. 1655 (1966 Act Creating the Department of Transportation).

Public Law 85-726, 72 Stat. 740, 49 USC. 1303 (Federal Aviation Act of 1958).

Public Law 75–706, 52 Stat. 973 USC. 1302 as amended (Civil Aeronautics Act of 1938).

REDFORD, E.S. (1961) 'Congress Passes the Federal Aviation Act of 1958' University of Alabama: University of Alabama Press Case Study 62.

SALACUSE, J.W. (1980) 'The Little Prince and the Businessman: Conflicts and Tensions in Public International Laws', *Journal of Air Law and Commerce* 45, 4: 807–44.

SIMONSON, G.R. (ed.) (1968) *The History of The American Aircraft Industry: Air Anthology*, Cambridge, MA: The MIT Press.

SMITH, C.R. (1971) *Safety in Air Transport Over the Years*, New York: Wings Club.

SMITH, H.L. (1965) *Airways: The History of Commercial Aviation in the US*, New York: Russell & Russell, Inc.

SOLBERG, C. (1979) *Conquest of the Skies: A History of Commercial Aviation in America*, Boston: Little, Brown and Company.

STRATZHEIM, M.H. (1969) *International Airline Industry*, Washington, DC: The Brookings Institute.

STROUD, J. (1971) *The World's Airliners*, London: The Bodley Head Ltd.

Survival in the Air Age: A Report by the President's Air Policy Commission (1948) Washington, DC: Government Printing Office.

THAYER, F.C., Jr. (1965) *Air Transport Policy and National Security: A Political Economic and Military Analysis*, Chapel Hill, NC: The University of North Carolina Press.

UNITED STATES DEPARTMENT OF TRANSPORTATION (1971) *Civil Aviation Research & Development Policy Study: A Historical Study Derived from Application of Technical Advances to Aviation*, Washington, DC: US Government Printing Office.

US International Aviation Policy at the Crossroads: A Study of Alternative Policies and Their Consequences, Volume I (1975) Boston: Harbridge

House for the Department of State, Department of Transportation, and President's Council on International Economic Policy.
WEBB, J.E. (1949) 'Department of State' *Air Affairs* (Autumn): 34–43.
WHITNAH, D.R. (1966) *Safer Skyways*, Ames: Iowa State University Press.

Industry Dynamics and Value

The ABCs of International Standardization (1981) New York: American National Standards Institute.
AECMA: Association Européenne des Constructeurs de Matériel Aérospatial (1982) Western Germany: Association Européenne des Constructeurs de Matériel Aérospatial.
Aerospace Industries Association Annual Report (1984) Washington, DC: Aerospace Industries Association of America.
AEROSPACE RESEARCH CENTER (1983a) *National Benefits of Aerospace Exports*, Washington, DC: Aerospace Industries Association of America.
—— (1983b) *The Challenge of Foreign Competition to the US Jet Transport Manufacturing Industry*, Washington, DC: Aerospace Industries Association of America.
—— (1985) '1985 Aerospace Year-End Review and Forecast: An Analysis', Washington, DC; Aerospace Industries Association of America.
AIA Aerospace Facts & Figures 1978/79 (1978) Washington, DC: Aerospace Industries Association of America.
AIA Aerospace Facts & Figures 1979/80 (1979) Washington, DC: Aerospace Industries Association of America.
AIA Aerospace Facts & Figures 1980/81 (1980) Washington, DC: Aerospace Industries Association of America.
AIA Aerospace Facts & Figures 1981/82 (1981) Washington, DC: Aerospace Industries Association of America.
AIA Aerospace Facts & Figures 1982/83 (1982) Washington, DC: Aerospace Industries Association of America.
AIA Aerospace Facts & Figures 1983/84 (1983) Washington, DC: Aerospace Inudtsries Association of America.
AIA Aerospace Facts & Figures 1984/85 (1984) Washington, DC: Aerospace Industries Association of America.
AIA Aerospace Facts & Figures 1985/86 (1985) Washington, DC: Aerospace Industries Association of America.
AIA Aerospace Facts & Figures 1986/87 (1986) Washington, DC: Aerospace Industries Association of America.
'Airbus: Still Flapping' (1984) *The Economist* 290, 7329 (18 February): 61.
'Aircraft Issues Buried in Stalled Talks' (1978) *Aviation Week and Space Technology* 109, 7 (14 August): 22–3.
'Airlines Approve Reorganizing IATA to Set Own Fares' (1978) *Wall Street Journal* (15 November): 14.
'Airlines at IATA Parley Decide to Seek Rise in Passenger Fares and Cargo Rates (1980) *Wall Street Journal* (6 December): 3.

Air Transport 1981: The Annual Report of the United States Scheduled Airline Industry (1981) (June) Washington, DC: Air Transport Association of America.

Air Transport 1982: The Annual Report of the United States Scheduled Airline Industry (1982) (June) Washington, DC: Air Transport Association of America.

Air Transport 1983: The Annual Report of the United States Scheduled Airline Industry (1983) (June) Washington, DC: Air Transport Association of America.

Air Transport 1984: The Annual Report of the United States Scheduled Airline Industry (1984) (June) Washington, DC: Air Transport Association of America.

Air Transport 1985: The Annual Report of the United States Scheduled Airline Industry (1985) (June) Washington, DC: Air Transport Association of America.

Air Transport 1986: The Annual Report of the United States Scheduled Airline Industry (1986) (June) Washington, DC: Air Transport Association of America.

Air Transport 1987: The Annual Report of the United States Scheduled Airline Industry (1987) (June) Washington, DC: Air Transport Association of America.

'Air Transport "Sacred Cow" Must Go, Skytrain Boss Says' (1979) *Rocky Mountain News* (23 0ctober): 32.

'Air Transport World 1983 Market Development Report' (1984) *Air Transport World* 21, 5 (May): 62–160.

American National Standards Institute 1981 Progress Report (1982) New York: American National Standards Institute.

'America's 1982 Foreign Trade Plummets into Deficit First Time in 3 Years; Jobless Claims Rise' (1983) *Rocky Mountain News* 330 (18 March): 30–1.

AMES, R.S. (1987) 'US Must Understand the Link Between R & D and the Economy', *Aviation Week and Space Technology* (12 October): 149–50.

AMIRAULT, J-M. (1980) *Requirements for Short/Medium Range Aircraft for the 1980s* G.2110/R1 (Second Edition, June) Brussels, Belgium: Association of European Airlines.

Association of European Airlines: Facts and Figures 1981 (1982) (April) Brussels, Belgium: Association of European Airlines.

BACHER, T.J. (1983) 'International Collaboration on Commercial Airplane Programs' Presentation at Conference Sponsored by Society of Japanese Aerospace Companies, Tokyo (29 March).

—— (1984) 'The Economics of the Commercial Aircraft Industry', Presentation at the Conference on The Role of South-East Asia in World Airline and Aerospace Development, Singapore (24–5 September), sponsored by the *Financial Times Limited*.

'BAe Makes its Case for A320 Participation' (1984) *Air Transport World* 21, 2 (February): 24–5.

'Basic Air-Fare Levels on International Runs are Adjusted by CAB' (1981) *Wall Street Journal* (1 September): 19.

'A Bitter Clash over Airlifters' (1982) *Business Week* (12 July): 91–2.

'Blowout Cancels Jetliner's Takeoff' (1984) *Denver Post* 93, 5 (6 August): 7A.

BLUESTONE, B., P. JORDAN, and M. SULLIVAN (1981) *Aircraft Industry Dynamics: An Analysis of Competition, Capital and Labor*, Boston, MA: Auburn House Publishing Company.

'Boeing, Japan Sign Work Share Pact for 7-7' (1984) *Aviation Week and Space Technology* (19 March): 32.

'Boeing Places Substantial Foreign Involvement in 7-7 Development' (1985) *Aviation Week and Space Technology* (3 June): 211–12.

BORETSKY, M. (1975) 'Trends in US Technology: A Political Economist's View', *American Scientist* 63 (January–February): 70-82.

BRANCKER, J.W.S. (1977) *IATA and What It Does*, Leyden: A A.W. Sijthoff.

'British Government Approves British Airways Merger Proposal' (1987) *Aviation Week and Space Technology* (16 November): 36.

'A British "No" Could Cripple Airbus Industries' (1984) *Business Week* (16 January): 39.

BROWN, D.A. (1982) 'European Aerospace Outlook: New Funding Sources Sought', *Aviation Week and Space Technology* 117, 10 (6 September): 80–5.

BROWN, D.A. (1978) 'Britain Rejects US Economy Fares' *Aviation Week and Space Technology* 108, 8 (20 February): 27.

BUERGENTHAL, T. (1969) *Law Making in the International Civil Aviation organization*, New York: Syracuse University Press.

BURCKHARDT, R. (1967) *The Federal Aviation Administration*, New York: Frederick A. Praeger.

'CAB Proposes Sharply Liberalized Charters' (1978) *Aviation Week and Space Technology* 108, 12 (20 March): 33.

'CAB Rejects IATA Request for Rate Boosts' (1979) *Wall Street Journal* (27 October): 8.

'CAB Softens Proposal to Quit Approving Fare-Setting Accords by IATA Members' (1979) *Wall Street Journal* (6 December): 3.

CARLEY, W.M. (1978) 'Boeing Waging Rare Attack on Politics of Airbus Industries in Selling Its A300', *Wall Street Journal* (13 April): 14.

—— (1979) 'Time of Turbulence: World's Airlines Face a Long Range Impact From DC10 Grounding', *Wall Street Journal* (18 June): 1+.

CARLEY, W.M. and D.P. GARINO (1979) 'Spreading Concern: DC10 Grounding Gives McDonnell Douglas Carriers New Problems', *Wall Street Journal* (8 June); 1+.

'The CF6' (1977) *Aviation Week and Space Technology* 107, 19 (7 November): 35.

CHAMBERS, A.B. (1980) 'Statement Before the Subcommittee on Aviation', Committee on Commerce, Science and Transportation, Washington, DC: US Senate.

CHAPLIN, J.C. (1982) 'What is JAR?', Presentation to the International Business Aviation Council Ltd, December.

COHEN, M.S. (1980a) 'Airbus, Airports and Competition: 1980', Statement by Civil Aeronautics Board Chairman to the Aero Club (28 October) Washington, DC.

—— (1980b) 'Airline Deregulation – It's Working', Press Release (24 October) Washington, DC: Civil Aeronautics Board.

COLEMAN, H.J. (1982) 'National Research Policy Aimed to Bolster Aviation', *Aviation Week and Space Technology* 117, 20 (15 November): 22–3+.

COMES, F.J. (1987) 'Widebody Wars: Airlines Decide "To Go For The Kill"' *Business Week* (6 July): 80–1.

Common European Procedures for the Authorisation of Category II and III Operations (1979) ECAC Document No. 17, Paris, France: European Civil Aviation Conference.

'Common Market Threats Force Carriers Toward Liberalization' (1987) *Aviation Week and Space Technology* (9 November): 146.

The Competitive Status of the US Civil Aviation Manufacturing Industry: A Study of the Influences of Technology in Determining International Industrial Competitive Advantage (1984). Prepared by the US Civil Aviation Manufacturing Industry Panel, Committee on Technology and International Economic and Trade Issues of the Office of the Foreign Secretary, National Academy of Engineering and the Commission on Engineering and Technical Systems, National Research Council, Frederick Seitz, Chairman and Lowell W. Steele, Rapporteur, Washington, DC: National Academy of Sciences.

'Continued International Cooperation in Aerospace: a Requirement for Success' (1987) Washington, DC: Aerospace Industries Association.

Convention on International Civil Aviation . . . the First 35 Years (1979) Montreal, Canada: International Civil Aviation Organisation.

'Council Decision of 15 December 1980 Adopting the Annual Report on the Economic Situation in the Community and Laying Down the Economic Policy Guidelines for 1981' (1980) (80/1265/EEC) *Official Journal of the European Communities* L375, 23 (31 December 1980): 17-31.

CULLEY, H. (ed) (1987) 'US Trade Policy', *GIST*, Washington, DC: Department of State.

CUTLER, M. (1978) Letter to John L. Burton, Chairman of the Subcommittee on Government Activities and Transportation of the House Committee on Government Operations (13 October) Washington, DC.

DALY, J.C. (Moderator) and A.V. CASEY, E.J. COLODNY, E.M. KENNEDY, L. MUSE, J ROBSON (1977) *Competition in the Airlines: What is in the Public Interest?* (Panel of 12 July) Washington, DC: American Enterprise Institute for Public Policy Research.

DeLAUER, R.D. (1982) 'The FY 1983 Department of Defense Program for Research, Development and Acquisition' (Statement by Undersecretary of Defense, Research and Engineering to the 97th Congress, Second Session) Washington, DC.

'Delta Airlines Says It Will Leave IATA Because CAB Rulings Weakened Group' (1978) *Wall Street Journal* (13 November): 19.

DEMISCH, W., C. DEMISCH, T. CONCERT (1984) 'The Jetliner Business' Special Report AE1991, Boston: The First Boston Corporation.

'Does The United States Support Its Commercial Transport Manufacturers Like Europe Supports Airlines? That's the European View . . . But

There is a Difference' (1987) Washington, DC: Aerospace Industries Association.
DOGANIS, R. (1984) *Flying off Course – The Economics of International Airlines*, London: George Allen & Unwin.
'The Dollar Effect of DC10 Inspection' (1979) *Business Week* (11 June): 36.
'Dollar Slide Seen Problem for Airbus' (1978) *Aviation Week and Space Technology* 108, 18 (1 May): 16.
DOTY, L. (1978a) 'US, Australian Policy Clash Expected', *Aviation Week and Space Technology* 108, 12 (20 March): 35–6.
—— (1978b) 'IATA Studies New Rate-Setting Methods', *Aviation Week and Space Technology* 108, 11 (13 March): 161–2.
—— (1978c) 'International Aviation Policy Shift Urged', *Aviation Week and Space Technology* 108, 6 (6 February): 36.
'A Drastic New Loss of Competitive Strength' (1980) *Business Week* (30 June): 58.
'Eastern Air Plans to Buy 19 Planes at French Firms' (1978) *Wall Street Journal* (7 April): 2.
'Eastern Changes Approach to Fleet Planning' (1982) *Aviation Week and Space Technology* 117, 10 (6 September): 66.
ECAC: European Civil Aviation Conference (1980) Paris, France: ECAC.
'Egypt's State Airline Reconfirms Its Order for Four DC10 Planes' (1979) *Wall Street Journal* (20 September): 12.
ELLINGSWORTH, R.K. (1974) 'Politics Clouds ICAO's Mission', *Aviation Week and Space Technology* 101, 21 (25 November): 20.
—— (1977) 'ICAO Parley Isolates Problems', *Aviation Week and Space Technology* 106, 18 (2 May): 32–3.
'European Aerospace Outlook: Bankers Revising Financing Methods' (1982) *Aviation Week and Space Technology* 117, 10 (6 September): 191–3.
'European Press Exports, Collaboration' (1978) *Aviation Week and Space Technology* 108, 11 (13 March): 51–3.
'European Report' (1984) *Air Transport World* 21, 6 (June): 4–9.
'Export-Import Bank Grows More Competitive' (1978) *Aviation Week and Space Technology* 109, 17 (23 October): 83.
'Farnborough Air Show: Flying Displays Showcase New Transports' (1982) *Aviation Week and Space Technology* 117, 11 (13 September): 18–20.
FEAVER, D.B. (1977) 'Landing Rights for the Concorde: A Diplomatic Dilemma', *Washington Post* (8 September): C9.
FEAZEL, M. (1984a) 'Deficits May Force "Buy Europe" Policy', *Aviation Week and Space Technology* 120, 7 (13 February): 135–40.
—— (1984b) 'Europe Moves to Ease Regulation', *Aviation Week and Space technology* 120, 5 (30 January): 28–9.
FELDMAN, J.M. (1984) 'The Children of Deregulation Have Brought Out the Best and Worst in Their Elders', *Air Transport World* 21, 1 (January): 52–7.
FUQUA, D. (1987) 'The Effect of Government Intervention on US Aerospace Industry International Competitiveness', Statement before the Subcommittee on Commerce, Consumer Protection and Competitiveness, Committee on Energy and Commerce, US House

of Representatives (23 June) Washington, DC: Aerospace Industries Association.

GOLICH, V.L. (1987) 'Market Induced State Cooperation: State Autonomy and the Internationalization of the Aerospace Industry', paper presented at the Annual Northeastern Meeting of the International Studies Association, Philadelphia, PA (November).

GREENHOUSE, S. (1987) 'Europeans Deregulate Air Travel', *The New York Times* (8 December): D1, D7.

GREGORY, W.H. (1978) 'British Back Pan Am L1011 Buy', *Aviation Week and Space Technology* 108, 15 (10 April): 21-2.

—— (1983) 'Military Sales to Extend Growth, Commercial Sector Remains Flat', *Aviation Week and Space Technology* 118, 11 (14 March): 8–9.

—— (1984) 'Editorial: Deregulation Goes International', *Aviation Week and Space Technology* 121, 1 (2 July): 11.

GRIFFITHS, D.R. (1978a) 'Eastern Orders Airbus, La Guardia Entry Eased', *Aviation Week and Space technology* 108, 15 (10 April): 27-8.

HAGGARTY, J.J. (1981) *Spinoff 1981: An Annual Report*, Washington, DC: NASA, Office of Space and Terrestrial Applications, Technology Transfer Division.

—— (1984) 'Aerospace Highlights 1983: Civil Aviation', *Aerospace* 22, 1 (Winter): 9–11.

HAGRUP, K. (1980) *Air Transportation in Europe*, Brussels, Belgium: Association of European Airlines.

HARR, K.G. (1984) 'The Aerospace Industry: A National Asset', *Aerospace* 22, 1 (Winter): 1.

HARTLEY, K. (1983) *NATO Arms Cooperation: A Study in Economics and Politics*, London: George Allen & Unwin.

HEARD, J. (1987) 'European Skies are Freer But No Friendlier', *Business Week* (21 December): 54.

IATA Backgrounder (1980) Montreal, Canada: International Air Transport Association.

'IATA Carriers to Seek 7% Fare Boosts; Airlines Set Cheaper US–London Rates' (1980) *Wall Street Journal* (29 July): 30.

'IATA's Fading Air-Fare Role' (1979) *Business Week* (26 November): 75–8.

'IATA To Boost Rates 7% to Compensate for Costlier Jet Fuel' (1979) *Wall Street Journal* (3 April): 6.

'IATA Votes to Urge Rate Rises Covering Passengers, Cargo' (1979) *Wall Street Journal* (10 October): 12.

ICAO Statistical Yearbook: Civil Aviation Statistics of the World – 1981 (1982) Doc. 9180/7. Montreal, Canada: International Civil Aviation Organisation.

ICAO Statistical Yearbook: Civil Aviation Statistics of the World – 1982 (1983) Doc. 9180/8. Montreal, Canada: International Civil Aviation Organisation.

'Icelandic Nonstop Bid Sparks Protest' (1979) *Aviation Week and Space Technology* 110, 23 (4 June): 27.

'Increased Flexibility Expected From New Regulation Proposal' (1982) *Aviation Week and Space Technology* 117, 23 (6 December): 56.

'Interagency Group on International Aviation' (1981) Washington, DC: Department of Transportation, FAA Office of Aviation Policy.
'Interest Rate Drop Eases Carrier Payments' (1982) *Aviation Week and Space Technology* 117, 17 (25 October): 31.
'International Airlines to Discuss Boosts in Fares; Carriers Cancel More Flights' (1979) *Wall Street Journal* (9 March): 6.
JANSSEN, R.F. (1975) 'The Airlines' Shaky Cartel', *Wall Street Journal* (15 October).
JOHNSEN, K. (1978a) 'Expansion of Exim Approved by Congress in Final Hours', *Aviation Week and Space Technology* 109, 17 (23 October): 30.
—— (1978b) 'Non-US Export Financing Scrutinized', *Aviation Week and Space Technology* 108, 14 (3 April): 59.
Joint Requirements for Emergency and Safety of Airborne Equipment (1979) ECAC.CECAC Document No. 18, Paris, France: European Civil Aviation Conference.
KAPLAN, D. (1979) 'Airline Deregulation Benefits Airlines and Public', Press Release, Washington, DC: Civil Aeronautics Board.
KARR, A.R. (1979b) 'CAB is Likely to Ease Plan for Ending Its Approval of International Air Transport Association Fare-Setting Pacts', *Wall Street Journal* (30 October): 10.
—— (1980) 'CAB Sheds Workers As It Prepares to Fly Slowly Into Sunset', *Wall Street Journal* (11 November): 1+.
KOLCUM, E.H. (1987) 'Commercial Air Transport Gains As Dollar Earner in South America', *Aviation Week and Space Technology (9 November): 133–5.*
KOZICHAROW, E. (1978) 'Aerospace Exports Surpass $9 Billion', *Aviation Week and Space Technology* 108, 11 (13 March): 55–9.
—— (1982) 'European Aerospace Outlook: Funding Spurs Growth in Italian Industry', *Aviation Week and Space Technology* 117, 10 (September): 183–6.
LAWLER, A.J. (1985) Presentation to the Air Transportation Research International Forum (12 June) Indianapolis, IN.
LEFER, H. (1978) 'New Airframe, Engine Designs Emphasize Reduced Maintenance', *Air Transport World* (September): 57–66.
LENOROVITZ, J.M. (1982) 'Airbus Near A320 Go-Ahead Decision', *Aviation Week and Space Technology* 117, 10 (6 September): 87–94.
LEVINE, M. (1986) 'Deregulation – Seven Years and Beyond', Presentation at the Federal Aviation Administration Annual Forecasting Meeting: Structural Changes in Aviation (27 February).
'Lockheed Takes the Lead' (1958) *Wall Street Journal* (1 April): 28.
LOPEZ, V. and L. YAGER (1987) 'An Aerospace Profile: The Industry's Role in the Economy; The Importance of R & D', *Facts and Perspectives* (April) Washington, DC: Aerospace Research Center.
LOWNDES, J.C. (1982) 'US Seeking Stronger Standards Role', *Aviation Week and Space Technology* 117, 8 (23 August): 25–6.
MANCHESTER, J.F. (1979) Statement to the Subcommittee on Government Activities and Transportation of the House Committee on Government Operations, Washington, DC, 18 June.

MANN, P. (1981a) 'Four Nations in Accord on Exports in Aircraft', *Aviation Week and Space Technology* 115, 6 (10 August): 25.
—— (1981b) 'Exim to Eliminate Credit for Older Transports', *Aviation Week and Space Technology* 115, 4 (2 July): 3.
—— (1982) 'US Exports Continue to Decline', *Aviation Week and Space Technology* 117, 10 (6 September): 51–2.
—— (1983) 'Aircraft Landing Demand Expected to Rise at Exim', *Aviation Week and Space Technology* 118, 11 (14 March): 196–8.
—— (1984) 'Civil Aircraft Exports Forecast to Drop $1 Billion', *Aviation Week and Space Technology* 120, 11 (12 March): 188–9.
'Masters of the Air' (1980) *Time* (7 April): 54.
Memorandum on ICAO (1978) (January) Montreal: International Civil Aviation Organisation.
MERCER, G. (ed.) (1980) *FAA Aviation Forecasts: Fiscal Years 1981–1982*, Washington, DC: Department of Transportation, FAA Office of Aviation Policy.
MONDOFF, K.F. (1987) 'European Air Transport Programs Prepare for Growth in Demand', *Aviation Week and Space Technology* (9 November): 71–4.
MORROCCO, J.D. (1987) 'Pentagon Reviews Trade Pacts to Standardize Joint Projects', *Aviation Week and Space Technology* (19 October): 45–9.
National Academy of Sciences, Transportation Research Board (1981), 'Assumptions and Issues Influencing the Future Growth of the Aviation Industry', *Transportation Research Circular* 230 (August): 1–50.
—— (1982) 'Uses and Misuses of Airline Traffic, Financial and Cost Data', *Transportation Research Circular* 239 (January): 2–44.
National Aeronautical R & D Goals: Technology for America's Future (1985) Washington, DC: Executive Office of the President, Office of Science and Technology Policy.
National Interest Aspects of the Private International Air Carrier System of the United States (1974) Washington, DC: Department of Commerce.
National Transportation Safety Board (1980) Washington, DC: National Transportation Safety Board.
'Nations Building Industrial Bases With Cooperative Production Efforts' (1985) *Aviation Week and Space Technology* (3 June) 323–7.
NEWHOUSE, J. (1982) *The Sporty Game: The High-Risk Competitive Business of Making and Selling Commercial Airliners*, New York: Alfred A. Knopf.
NEUMAN, S. (1984) 'International Stratification and Third World Military Industries', *International Organization* (Winter): 167–97.
'New Pact Counters Bermuda 2' (1978) *Aviation Week and Space Technology* 108, 13 (27 March): 25–6.
'News Briefs: Federal Express Keeps DC10 Alive' (1984) *Air Line Pilot* 53, 7 (July): 33.
'News Digest' (1982) *Aviation Week and Space Technology* 116, 10 (8 March): 280.
O'CONNER, W.E. (1971) *Economic Regulation of the World's Airlines*, New York: Praeger Publishers.

—— (1978) *An Introduction to Airline Economics*, New York: Praeger Publishers.

O'DONNELL, J.J. (1981) National News Media Press Conference: Opening Statement, Questions and Answers (19 August) Washington, DC: Air Line Pilots Association.

—— (1982) *President's Report*, Washington, DC: Air Line Pilots Association.

'Offer of Airbus to United Draws Treasury Scrutiny' (1978) *Aviation Week and Space Technology* 109, 4 (24 July): 18.

O'LONE, R.G. (1978) 'New Aircraft Program Timing', *Aviation Week and Space Technology* 108, 10 (6 March): 23–4.

—— (1983a) 'US Manufacturers Project Turnaround', *Aviation Week and Space Technology* 118, 11 (14 March): 167–75.

—— (1983b) 'Boeing Studies Increased Joint Projects', *Aviation Week and Space Technology* (30 May): 222–7.

OTT, J. (1982a) 'Delta Cuts Fares on Short-Haul Flights', *Aviation Week and Space Technology* 117, 25 (20 December): 31–3.

—— (1982b) 'Northwest Delays Aircraft Replacement Until 1985', *Aviation Week and Space Technology* 117, 20 (15 November): 30.

—— (1984) 'US Stiffens Negotiating Stance', *Aviation Week and Space Technology* 120, 15 (9 April): 28–30.

'Pan-Am/Scandia Blocked Space Agreement Approved Under State Pressure' (1979) *Aviation Daily* 241, 20 (29 January): 156.

'Paris Air Show Offers Venue for Cooperative Agreements' (1985) *Aviation Week and Space Technology* (3 June): 116–7.

PASCALE, C. (1987) '1986–87 Aerospace Industry Employment Survey' (October 5) 87–51; 11–01; 86–11/11–01, Washington, DC: Aerospace Industries Association.

POLLOCK, A.M. (1984) 'US and Ten Other Nations Approve Agreement to Give Airlines More Fare Flexibility in North Atlantic Markets', Press Release, Civil Aeronautics Board (3 May) Washington, DC.

PREBLE, C. (1984) 'FAA Noise Rules Benefiting Nacelle Design Companies', *Aviation Week and Space Technology* 121, 4 (23 July): 39–47.

'Privatization Sparks Scramble for Competitive Edge in England' (1987) *Aviation Week and Space Technology* (9 November): 128–33.

'Procedures Concerning Supplier Audits and Implementation of Bilateral Agreement' (1982) DOT FAA Advisory Circular No. 21–AA, *Federal register* 27, 45 (8 March): 9948–52.

REED, A. (1984) 'BAe Makes Its Case for A320 Participation', *Air Transport World* 2: 24–5.

Review of the Economic Situation of Air Transport 1969–1979 (1980) Circular 158-AT/57 (August) Montreal: International Civil Aviation Organisation.

REVZIN, P. (1979) 'High European Air Fares Could Descend as Independent Carriers Seek Rate Cuts', *Wall Street Journal* (27 November): 10.

ROPELEWSKI, R.R. (1978) 'Europeans Liberalize Fares, Charters', *Aviation Week and Space Technology* 108, 13 (27 March): 27–8.

—— (1981) 'Heavy Losses Cited in Decision to Terminate L10lls', *Aviation Week and Space Technology* 115, 24 (14 December): 26–9.

—— (1983) 'New DC9 Model to Offer Greater Range', *Aviation Week and Space Technology* 118, 6 (February 7): 30–1.

RUDOLPH, B. (1987) 'Trouble on the Horizon: Europe's Airlines Threaten the Dominance of US Jet Builders', *Time* (11 May): 50–1.

'Scuttling of L1011 May Cost Rolls Revenue of $450 Million' (1982) *Aviation Week and Space Technology* 117, 11 (13 September): 96.

SHANE, J.N. (1988) 'Challenges in International Civil Aviation Negotiations', United States Department of State, Public Information Series, Washington, DC: Bureau of Public Affairs.

SMITH, B.A. (1982) 'American to Lease New DC9-80s', *Aviation Week and Space Technology* 117, 114 (October 4): 28–29.

STUART, A. (1980) 'The Airlines are Flying in a Fog', *Fortune* (20 October): 50–5.

SULLIVAN, S. (1984) 'The Decline of Europe', *Newsweek* 103, 15 (9 April): 44–56.

'Super 80 Crew Size Triggers Debate' (1979) *Aviation Week and Space Technology* (29 November): 27.

'Sweeteners for Foreign Aircraft Sales' (1979) *Industry Week* (28 May): 87.

'757 Effort Involves 1 300 Companies' (1982) *Aviation Week and Space Technology* (22 February): 44–5.

TANEJA, N.K. (1970) *US International Aviation Policy*, Lexington, MA: Lexington Books.

—— (1976) *The Commercial Airline Industry*, Lexington, MA: Lexington Books.

'The Technology Edge' (1987) *Flight International* (14 February): 1.

'Towards Agreement on a Joint Airlines Project' (1967) *Interavia* 22 (March): 328.

'Trans World Intends to Lease 15 DC9-80s' (1982) *Aviation Week and Space Technology* 117, 16 (18 October): 22.

'TWA Board Approves DC9-80 Leasing' (1982) *Aviation Week and Space Technology* 117, 18 (1 November): 36.

'UK Firm Uses US Experience to Break Into Other Markets' (1977) *Aviation Week and Space Technology* 106, 23 (6 June): 159–60.

'United Chief Hits Foreign Financing' (1978) *Aviation Week and Space Technology* 108, 18 (1 May): 24.

UNITED NATIONS (1988) *1986 International Trade Statistics Yearbook*, New York.

UNITED STATES BUREAU OF THE CENSUS (1982) *Statistical Abstract of the United States: 1982–83* (103d Edition) Washington, DC: US Government Printing Office.

UNITED STATES CIVIL AERONAUTICS BOARD (1970) *Aeronautical Statutes and Related Materials* (Revised Edition – 1 June) Department of Transportation Act of 15 October 1966, Section 2, Washington, DC: US Government Printing Office.

—— (1978) *Summary of the Airline Deregulation Act of 1978*, Public Law 95–504, Washington, DC: US Government Printing Office.

UNITED STATES DEPARTMENT OF COMMERCE (1980) *Agreement*

on Technical Barriers to Trade: A Descriptive Summary, (June), Washington, DC: International Trade Administration, Office of Trade Policy.

—— (1983) *An Assessment of US Competitiveness in High Technology Industries*, Washington, DC: Department of Commerce, International Trade Administration.

—— (1988) *US Industrial Outlook*, Washington, DC: US Government Printing Office.

UNITED STATES DEPARTMENT OF STATE (1972a) *Certificates of Airworthiness for Imported Aircraft Agreement Between the United States of America and the United Kingdom of Great Britain and Northern Ireland, December 28*, Washington, DC: Government Printing Office.

—— (1972b) *Memorandum of Law – Subject: Circular 175 – Request for Authorization to Negotiate and Sign New Bilateral Airworthiness Agreements, December 5*, Washington, DC: Government Printing Office.

UNITED STATES DEPARTMENT OF TRANSPORTATION (1974a) *Federal Aviation Regulations, Part 33, Airworthiness Standards: Aircraft Engines, August*, Washington, DC: Federal Aviation Administration.

—— (1974b) *Federal Aviation Regulations, Part 21, Certification Procedures for Products and Parts, May*, Washington, DC: Federal Aviation Administration.

—— (1979) *Federal Aviation Administration Organization Manual*, DOT Order 1100.148, Appendix 1, 14 June, Washington, DC: Office of Management Planning.

—— (1981) *International Aviation Programs*, Order 1240.9, 2 July, Washington, DC: Federal Aviation Administration.

—— (1982a)*Aircraft Certification Functions, History, Procedures – Handout for Presentation to National Aircraft Certification Organization, March 18*, Washington, DC: Federal Aviation Administration.

—— (1982b) *Interim Operating Procedures for Aircraft Certification Directorates*, Notice N8100.5, 10 February, Washington, DC: Federal Aviation Administration Office of Airworthiness.

—— (1982c) *Aircraft Certification Directorates*, Order 8000.51, 1 February, Washington, DC: Federal Aviation Administration Office of Airworthiness.

—— (1986) *FAA Aviation Forecasts: Fiscal Years 1986–1987*, FAA-APO-86a, Washington, DC: Federal Aviation Administration.

UNITED STATES HOUSE OF REPRESENTATIVES (1974a) *International Air Transportation Fair Competitive Practices Act of 1974: Report Together with Dissenting View*, Washington, DC: US Government Printing Office.

—— (1974b) *International Air Transportation Competition: Hearings before the Committee on Interstate and Foreign Commerce and the Subcommittee on Transportation and Aeronautics*, Washington, DC: US Government Printing Office.

—— (1976a) *Assessing the New Political Trend, Dublin, 1976*, Report on the Ninth Meeting of Members of Congress and of the European Parliament, 21–3 April, Washington, DC: US Government Printing Office.

—— (1976b) *The Future of Aviation. Volume I: Report prepared by the Subcommittee on Aviation and Transportation Research and Development of the Committee on Science and Technology*, Washington, DC: US Government Printing Office.

—— (1976c) *The Future of Aviation. Volume II: A Compilation of Papers prepared by the Subcommittee on Aviation and Transportation R & D of the Committee on Science and Technology*, Washington, DC: US Government Printing Office.

—— (1976d) *The Future of Aviation: Hearings before the Subcommittee on Aviation and Transportation R & D of the Committee on Science and Technology*, Washington, DC: US Government Printing Office.

UNITED STATES SENATE (1978) *International Aviation Hearings before the Subcommittee on Aviation of the Committee on Commerce, Science and Transportation*, 95th Congress, 1st Session, Washington, DC: US Government Printing Office.

'US Airlines Criticize Icelandic Exemption Bid' (1979) *Aviation Daily* 246, 15 (26 November): 117.

'US Airlines Granted Authority to Serve Korea' (1979) *Interavia Air Letter* 9286 (2 July): 4.

'US, Japan Bilateral Talks Remain Stalemated' (1978) *Aviation Week and Space Technology* 108, 14 (3 April): 34.

'US-Japan Negotiations Break Down' (1982) *Aviation Week and Space Technology*, 116, 12 (22 March): 32.

'US/Japan Talks Appear Stalemated' (1982) *Aviation Week and Space Technology*, 116, 12 (22 March): 32.

'US Lifts Ban on Jet Engine Venture with France's SNECMA' (1973) *New York Times* (23 June): 37.

'US Makes Counter Offer on Icelandic Request' (1979) *Aviation Daily* 244, 38 (23 August): 299.

'US Officials Say Negotiations Progressing on Export Credits' (1981) *Aviation Week and Space Technology*, 115, 3 (20 July): 86–7.

'US Officials See End to Predatory Financing' (1980) *Aviation Week and Space Technology* 112, 21 (26 May): 23.

'US Reassesses Role in ICAO After Vienna Assembly Walkout' (1971) *Aviation Daily* 196, 21 (30 July): 164.

US Service Industries in the World Markets: Current Problems and Future Policy Development (1976) (December) Washington, DC: Department of Commerce.

'Washington Roundup: Exim Seeks Raise' (1982) *Aviation Week and Space Technology* 117, 21 (22 November): 15.

'Washington Roundup: Financing Frustration' (1980) *Aviation Week and Space technology* 113, 25 (29 December): 13.

'Washington Roundup: Financing Flap' (1980) *Aviation Week and Space Space Technology* 112, 24 (16 June): 25.

'Washington Roundup: Predatory Financing' (1980) *Aviation Week and Space Technology* 114, 19 (11 May): 5.

WELLES, N. (1980) 'The 50 Leading Exporters', *Fortune* (22 September): 116.

WELLING, B. (1982) 'The Airlines Dilemma: No Cash to Buy Fuel-Efficient Jets', *Business Week* 2758 (27 September): 65.

WHETMORE, W.C. (1978) 'Equipment Financing Outlook Brighter',

Aviation Week and Space Technology 109, 17 (23 October): 83–94.
WILSON, T.A. (1978) 'American Sanctions Vs Exports', *Aviation Week and Space Technology* 109, 18 (30 October): 9.
YAGER, L. (1987) 'US Aerospace Trade Continues to Grow: Share of World Market Declines', *Facts and Perspectives* (September) Washington, DC: Aerospace Research Center.

Theory

ALLISON, G.T. (1971) *Essence of Decision* Boston: Little, Brown & Co.
ARONSON, J.D. (1977) *Money and Power: Banks and the World Monetary System* (Sage Library of Social Research, vol. 66) Beverly Hills: Sage Library of Social Research.
AXELROD, R. and R. KEOHANE (1984) 'Achieving Cooperation Under Anarchy: Strategies and Institutions', *World Politics* 38 (October): 226–54.
BECKER, G.S. (1986) 'Why Public Enterprises Belong in Private Hands', *Business Week* (24 February): 20.
BENDUR, J. and D. MOOKHERJEE (1987) 'Institutional Structure and the Logic of Ongoing Collective Action', *American Political Science Review*, 81, 1 (March): 129–54.
BORRUS, M., J.R. MILLSTEIN and J. ZYSMAN (1983) 'Trade and Development in the Semiconductor Industry: Japanese Challenge and American Response' in Zysman, J and L. Tyson (eds) *American Industry in International Competition: Government Politics and Corporate Strategies*, Ithaca, NY: Cornell University Press.
BORRUS, A., F.J. COMES, R. FLY, F.A. MILLER and J. TEMPLEMAN (1987) 'Limping Into Venice – Domestic Political Woes May Sink the Summit', *Business Week 3002 (8 June): 34–5.*
BULL, H. (1977) *The Anarchical Society*, London: The Macmillan Press.
BROWN, S., N.W. CORNELL, L.L. FABIAN and E.B. WEISS (1977) *Regimes for the Ocean, Outer Space, and Weather*, Washington, DC: The Brookings Institute.
BURNHAM, W.D. (1982) 'The Constitution, Capitalism and the Need for Rationalized Regulation', in R.A. Goldwin and W.A. Shambra (eds). *How Capitalistic is the Constitution*, pp. 75–105, Washington, DC: American Enterprise Institute for Public Policy Research.
CLAUDE, I. (1964) *Swords into Plowshares: The Problems and Progress of International Organization*, New York: Random House.
COHEN, B.J. (1977) *Organizing the World's Money: The Political Economy of International Monetary Relations*, New York: Basic Books Inc., Publishers.
—— (1982) 'Balance-of-Payments Financing: Evolution of a Regime', *International Organization* 36, 2 (Spring): 457–78.
CONYBEARE, J.A.C. (1982) 'The Rent Seeking State and Revenue Diversification', *World Politics*, 35 (October): 25–42.
—— (1984) 'Public Goods, Prisoners' Dilemmas and the International

Political Economy', *International Studies Quarterly* 28, 1 (March): 5–22.

COWHEY, P.F. and E. Long (1983) 'Testing Theories of Regime Changes: Hegemonic Decline or Surplus Capacity?', *International Organization* 37, 2 (Spring): 157–88.

COX, R. and H. JACOBSON (eds) (1973) *The Anatomy of Influence: Decision-Making in International Organizations*, New Haven, CT: Yale University Press.

CRAWFORD, B. and S. LENWAY (1985) 'Decision Modes and International Regime Change: Western Collaboration on East-West Trade', *World Politics* 37, (3): 375–402.

DAHL, R. (1963) *Modern Political Analysis*, Englewood Cliffs, NJ: Prentice-Hall.

DIXIT, A.K. and A.S. KYLE (1985) 'The Use of Protection and Subsidies for Entry Promotion and Deterrence', *The American Economic Review* 75 (March): 139–52.

DOUGLAS, G.K. (ed) (1979) *The New Independence: The European Community and the United States*, Lexington, MA: Lexington Books.

EASTON, D. (1965) *A Systems Analysis of Political Life*, New York: John Wiley and Sons, Inc.

ECKSTEIN, H. (1975) 'Case Study and Theory in Political Science', in F. Greenstein and N. Polsley (eds) *Handbook of Political Science*, vol. 7, pp. 79–137, 8 vols., Reading, MA: Addison-Wesley.

EDMONDS, M. (1978) 'National Investment, Policy Options and the International Aerospace Market', paper delivered at Columbia-CUNY Center for European Studies Annual International Political Economy Conference.

FALK, R. (1974)'The Logic of State Sovereignty Versus the Requirements of World Order', *1973 Yearbook of World Affairs*, London: Royal Institute of International Affairs.

—— (1975) *A Study of Future Worlds*, New York: Free Press.

FEIGENBAUM, H.B. (1982) 'Public Enterprise in Comparative Perspective', *Comparative Politics* (October): 101–22.

—— (1985) *The Politics of Public Enterprise: Oil and the French State*, Princeton, NJ: Princeton University Press.

FELD, W. (1979) 'Implementation of the European Community's Common Agricultural Policy: Expectations, Fears, Failures', *International Organization* 33, 3 (Summer): 335–63.

FRIEDHEIM, R.I., J.B. KADANE, K. GOUNDREAU and W.J. DURCH (draft) 'Forecasting Multilateral Negotiations'.

FROHLICH, N., J.A. OPPENHEIMER and O. YOUNG (1971) *Political Leadership and Collective Goods*, Princeton: Princeton University Press.

GADDIS, J.L. (1972) *The United States and the Origins of the Cold War*, New York: Columbia University Press.

GIDWITZ, B. (1980) *The Politics of International Air Transport*, Lexington, MA: Lexington Books.

GILPIN, R. (1975) *US Power and the Multinational Corporation: The Political Economy of Foreign Direct Investment*, New York: Basic Books, Inc.

—— (1977) 'Economic Interdependence and National Security in Historical

Perspective', in K. Knorr and F.N. Trager (eds) *Economic Issues and National Security*, Lawrence, KS: Allen Press, Inc.

—— (1981) *War and Change in International Politics*, New York: Cambridge University Press.

GOEDHIUS, D. (1947) *Idea and Interest in International Aviation*, The Hague: Martinus Nijhoff.

GOLICH, V.L. (1984) *The Politics and Economics of the International Commercial Aviation Safety Regime*, Dissertation for the School of International Relations at the University of Southern California.

GORDON, R.J. (1969) '$45 Billion of US Private Investment Has Been Mislaid', *American Economic Review* (June): 224.

GOWA, J. (1984) 'Hegemons, IOs and Markets: The Case of the Substitution Account', *International Organization* 38 (Autumn): 661–83.

—(1985) 'Subsidizing American Corporate Expansion Abroad: Pitfalls in the Analysis of Public and Private Power', *World Politics* 37, 2 (January): 180–203.

—— (1986) 'Anarchy, Egoism and Third Images: The Evolution of Cooperation and International Relations', *International Organization* 40, 1 (Winter): 167–86.

GROOM, A.J.R. and P. TAYLOR (eds) (1975) *Functionalism: Theory and Practice in International Relations*, London: University of London Press.

HAAS, E.B. (1964) *Beyond the Nation-State: Functionalism and International Organization*, Stanford, CA: Stanford University Press.

—— (1975) 'Is there a Hole in the Whole? Knowledge, Technology, Interdependence and the Construction of International Regimes', *International Organization* 29 (Summer): 827–92.

—— (1980a) 'Turbulent Fields and the Theory of Regional Integration', *International Organization* 30, 2 (Spring): 173–212.

—— (1980b) 'Why Collaborate? Issue-Linkage and International Regimes', *World Politics* 32 (April): 357–405.

—— (1982) 'Words Can Hurt You or Who Said What to Whom About Regimes', *International Organization* 36, 2 (Spring): 207–44.

HAAS, E.B., M.P. WILLIAMS and D. BABAI (1977) *Scientists and World Order: The Uses of Technical Knowledge in International Organizations*, Berkeley: University of California Press.

HAGGARD, S. and B.A. SIMMONS (1987) 'Theories of International Regimes', *International Organization* 491–517.

HARDIN, G. (1968) 'The Tragedy of the Commons', *Science* 162 (13 December): 1243–8.

HARDIN, R. (1971) 'Collective Action as an Agreeable n-Prisoners' Dilemma', *Behavioral Science* 16: 472–81.

HARRISON, J.R. and R.J. KOLCZYNSKI (1978) 'Government Liability for Certification of Aircraft?', *Journal of Air Law and Commerce* 44, 1: 23–45.

HART, J.A. (1986) 'Polarity, Hegemony and the Distribution of Power', paper prepared for delivery at the annual meeting of the American Political Science Association, Washington, DC: 28–31 August.

HART, J.A. and P.F. COWHEY (1977) 'Theories of Collective Goods Reexamined', *Western Political Quarterly* 30, 3: 351–62.

HOCHMUTH, M.S. (1974) 'Aerospace' in R. Vernon (ed) *Big Business and the State: Changing Relations in Western Europe*, Cambridge, MA: Harvard University Press.

HOLLAND, S. (1974) 'Europe's New Public Enterprises', in R. Vernon (ed) *Big Business and the State: Changing Relations in Western Europe*, Cambridge, MA: Harvard University Press.

HOLSTI, O.R., R.M. SIVERSON and A.L. GEORGE (1980) *Change in the International System*, Boulder, CO: Westview Press.

HOPKINS, R.F. (1976) 'The International Role of "Domestic" Bureaucracy', *International Organization* 30 (Summer): 405–32.

IKENBERRY, G.J. (1986) 'The State and Strategies of International Adjustment', *World Politics* 39, 1 (October): 53–77.

JACQUEMIN, A. (1983) 'Industrial Policies and the City', in P. Coffey (ed) *Main Economic Policy Areas in the EEC*, The Hague: Martinus Nijhoff Publishers.

JERVIS, R. (1982) 'Security Regimes', *International Organization* 36, 2 (Spring): 357–87.

—— (1983) 'Perception and Misperception: The Spiral of International Insecurity', in W.C. Olson, D.S. McClellan, F.A. Sonderman (eds) *The Theory and Practice of International Relations*, pp. 200–5, Sixth Edition, Englewood Cliffs, NJ: Prentice Hall.

—— (1988) 'Realism, Game Theory and Cooperation', *World Politics* 60, 3 (April): 317–49.

JÖNSSON, C. (1981) 'Sphere of Flying: The Politics of International Aviation', *International Organization* 35, 2 (Spring): 273–302.

—— (1986) 'Do Transnational Networks Matter?', prepared for presentation at the 27th annual convention of the International Studies Association, 25–9 March 1986, Anaheim, CA.

—— (1987a) *International Aviation and the Politics of Regime Change*, New York: St. Martin's Press.

—— (1987b) 'Interorganizational Relations and Regime Change: Findings for International Aviation', prepared for presentation at the 28th annual convention of the International Studies Association, 15–8 April 1987, Washington, DC.

KALIJARVI, T.V. (1957) 'The Paradox of Foreign Economic Policy', *Department of State Bulletin* (24 June): 36–78.

KATZENSTEIN, P.J. (1976) 'International Relations and Domestic Structures: Foreign Economic Policies of Advanced Industrial States', *International Organization* (Winter): 1–41.

—— (1985) *Small States in World Markets*, Ithaca, NY: Cornell University Press.

KEOHANE, R.O. (1975) 'International Organization and the Crisis of Interdependence', *International Organization* 29 (Spring): 357–65.

—— (1979) 'US Foreign Economic Policy Toward Other Advanced Capitalistic States', in K.A. Oye, D. Rothschild, R.J. Lieber (eds) *Eagle Entangled: US Foreign Policy in a Complex World*, pp. 91–122, New York: Longman, Inc.

—— (1980) 'The Theory of Hegemonic Stability and Changes', in O.R. Holsti, R.M. Siverson, A.L. George (eds) *Change in the International System*, pp. 131–62, Boulder, CO: Westview Press.

—— (1982a) 'Hegemonic Leadership and US Foreign Economic Policy in the "Long Decade" of the 1950s', in W. Avery and D.P. Rapkin (eds) *America in a Changing World Political Economy*, pp. 49– 76, New York: Longman, Inc.

—— (1982b)'Inflation and the Decline of American Power', in R. Lombra and W. Witte (eds) *The Political Economy of International and Domestic Monetary Relations*, pp. 7–37, Ames: Iowa State University.

—— (1982c) 'The Demand for International Regimes', *International Organization* 36, 21 (Spring): 325–56.

—— (1984) *After Hegemony: Cooperation and Discord in the World Political Economy*, Princeton, NJ: Princeton University Press.

—— (1986) 'Reciprocity in International Relations', *International Organization*, 40, 1 (Winter): 1–27.

KEOHANE, R.O. and J. NYE (1972) *Transnational Relations and World Politics*, Cambridge, MA: Harvard University Press.

—— (1974) 'Transgovernmental Relations and International Organizations', *World Politics* (October): 39–62.

—— (1977) *Power and Interdependence: World Politics in Transition*, Boston: Little, Brown and Company.

—— (1987) '*Power and Interdependence* Revisited', *International Organization* 41, 4 (Autumn): 725–53.

KINDLEBERGER, C.P. (1973) *The World in Depression 1929–1939*, Berkeley: University of California Press.

—— (1978a) *Manias, Panics and Crashes*, New York: Basic Books.

—— (1978b) 'Government and International Trade', Princeton Essays in International Finance, No. 129 (July) Camden NJ: International Finance Section, Princeton University.

—— (1986) 'Hierarchy Versus Inertial Cooperation', *International Organization* 40, 4 (Autumn): 841–7.

KRASNER, S.D. (1976) 'State Power and the Structure of International Trade', *World Politics* (April): 317–47.

—— (1982a) 'Structural Causes and Regime Consequences: Regimes as Intervening Variables' *International Organization* 36, 2 (Spring): 185–205.

—— (1982b) 'Regimes and the Limits of Realism: Regimes as Autonomous Variables' *International Organization* 36, 2 (Spring): 497–510.

—— (1985) *Structural Conflict: The Third World Against Global Liberalism*, Berkeley: University of California Press.

KRATOCHWIL, F. and J.G. RUGGIE (1986) 'International Organization: A State of the Art on An Art of the State', *International Organization*, 40, 4 (Autumn): 753–75.

KUTTNER, R. (1987) 'The Prospect for a Global Summit: Political Gridlock', *Business Week* 3002 (8 June): 22.

LAKE, D. (1983) 'International Economic Structure and American Foreign Economic Policy', 1887–1934, *World Politics* 35 (July): 517–43.

LASSWELL, H. (1965) *Politics: Who Gets What, When, How*, New York: Meridian Books.

LEIVE, D.M. (1976) *International Regulatory Regimes – Volumes I and II*, Lexington, MA: Lexington Books.

LIJPHART, A. (1971) 'Comparative Politics and the Comparative Method', *American Political Science Review* 65, 3 (September): 682–93.

LINDBLOM. C.E. (1977) *Politics and Markets: The World's Political–Economic Systems*, New York: Basic Books Inc.

LIPSON, C. (1982) 'The Transformation of Trade: The Sources and Effects of Regime Change', *International Organization* 36, 2 (Spring): 417–56.

McDONALD, F. (1982) 'The Constitution and Hamiltonian Capitalism', in R.A. Goldwin and W.A. Shambra (eds) *How Capitalistic is the Constitution?*, pp. 49–74, Washington, DC: American Enterprise Institute for Public Policy Research.

McKEOWN, T.J. (1983) 'Hegemonic Stability Theory and 19th Century Tariff Levels in Europe', *International Organization* 37, 1 (Winter): 73–91.

—— (1986) 'The Limitations of "Structural" Theories of Commercial Policy', *International Organization* 40, 1 (Winter): 43–64.

M'GONIGLE, R.M. and M.W. ZACHER (1980) *Pollution, Politics and International Law: Tankers at Sea*, Berkeley: University of California Press.

MILLER, S. (1982) 'The Constitution and the Spirit of Commerce', in R.A. Goldwin and W.A. Shambra (eds) *How Capitalistic is the Constitution?*, pp. 148–69, Washington, DC: American Enterprise Institute for Public Policy Research.

MITRANY, D. (1943) *A Working Peace System*, London: Royal Institute of International Affairs.

—— (1948) 'The Functionalist Approach to World Organization' *International Affairs* 24 (July): 350–63.

MODELSKI, G. (1972) *Principles of World Politics*, New York: The Free Press.

—— (1978) 'The Long Cycle of Global Politics and the Nation–State', *Comparative Studies in Society and History* 20, 2 (April): 214–35.

MOE, T. (1980) *The Organization of Interests: Incentives and the Internal Dynamics of Political Interest Groups*, Chicago: University of Chicago Press.

MORGENTHAU, H.J. (1960) *Politics Among Nations: The Struggle for Power and Peace*, New York: Alfred A. Knopf.

MORSE, E.L. (1977) 'Managing International Commons', *Journal of International Affairs* 31, 1: 1–21.

—— (1982) 'Transnational Economic Processes', in R. Keohane and J. Nye (eds) *Transnational Relations and World Politics*, pp.23– 48, Cambridge, MA: Harvard University Press.

NORTH, D.C. (1981) *Structure and Change in Economic History*, NY: Norton.

NORTHRUP, H.R. and R.L. ROWAN (1979) *Multinational Collective Bargaining Attempts: The Record, the Cases and the Prospects*, Multinational Industrial Relations Series No. 6, Philadelphia, PA: Industrial Research Unit, The Wheaton School, University of Pennsylvania.

NYE, J. (1976) 'Independence and Interdependence', *Foreign Policy* 22 (Spring): 130–61.

—— (1988) 'Neorealism and Neoliberalism' *World Politics* 60, 2 (January): 235–52.

ODELL, J.S. (1982) *US International Monetary Policy: Markets, Power and Ideas as Sources of Change*, Princeton: Princeton University Press.

OLSON, M. (1965) *The Logic of Collective Action: Public Goods and the Theory of Groups*, Cambridge, MA: Harvard University Press, Harvard Economic Studies Volume 124.

OYE, K.A. (1979) 'The Domain of Choice: International Constraints and the Carter Administration', in K.A. Oye, D. Rothchild and R.J. Lieber (eds) *Eagle Entangled: US Foreign Policy in a Complex World*, pp. 3–33, New York: Longman, Inc.

—— (1985) 'Explaining Cooperation Under Anarchy: Hypotheses and Strategies', *World Politics* 38: 1–24.

PAARLBERG, R.L. (1976) 'Domesticating Global Management', *Foreign Affairs* 54 (April): 563–76.

PHILLIPS, A. (1971) *Technology and Market Structure: A Study of the Aircraft Industry*, Lexington, MA: Heath Lexington Books.

PUCHALA, D.J. and R.F. HOPKINS (1982) 'International Regimes: Lessons from Inductive Analysis', *International Organization* 36, 2 (Spring): 245–76.

ROGOWSKI, R. (1983) 'Structure, Growth and Power: Three Rationalist Accounts', *International Organization* 37, 4 (Autumn): 713–38.

ROSECRANCE, R. (1981) 'International Theory Revisited', *International Organization* 35, 4 (Autumn): 691–713.

ROSECRANCE, R., A. ALEXANDROFF, W. KOEHLER, J. KROLL, S. LAQUER and J. STOCKER (1977) 'Whither Interdependence?', *International Organization* (Summer): 425–45.

ROSECRANCE, R. and A. STEIN (1973) 'Interdependence: Myth or Reality', *World Politics* 26 (October): 1–27.

ROSENAU, J.N. (1986) 'Before Cooperation: Hegemons Regimes, and Habit-Driven Actors in World Politics', *International Organization* 40, 4 (Autumn): 849–94.

RUGGIE, J.G. (1972) 'Collective Goods and Future International Collaboration', *American Political Science Review* 66 (September): 874–93.

—— (1975) 'International Responses to Technology: Concepts and Trends' *International Organization* 29 (Summer): 557–83.

—— (1982) 'International Regimes, Transactions, and Change: Embedded Liberalism in the Postwar Economic Order', *International Organization* 36, 2 (Spring): 379–416.

RUSSETT, B. (1985) 'The Mysterious Case of Vanishing Hegemony; or Is Mark Twain Really Dead?', *International Organization* 39, 2 (Spring): 207–31.

SCHELLING, T. (1963) *The Strategy of Conflict*, New York: Oxford University Press.

SEWELL, J.P. (1966) *Functionalism and World Politics: A Study Based on United Nations Programs Financing Development*, Princeton, NJ: Princeton University Press.

SHEPHERD, G. and F. DUCHENE (1983) 'Introduction: Industrial Change and Intervention in Western Europe', in G. Shepherd, F. Duchene, C. Saunder (eds) *Europe's Industries: Public and Private Strategies for Change*, Ithaca, NY: Cornell University Press.

SIEGEN, B.H. (1982 'The Constitution and the Protection of Capitalism', in R.A. Goldwin and W.A. Shambra (eds) *How Capitalistic is the Constitution?*, pp. 106–26, Washington, DC: American Enterprise Institute for Public Policy Research.

SKOLNIKOFF, E. (1972) *The International Imperatives of Technology*, Berkeley, CA: Institute of International Studies.

SMELSER, N.J. (1973) 'The Methodology of Comparative Analysis', in D.P. Warwick and S. Osherson (eds) *Comparative Research Methods* pp. 42–86, Englewood Cliffs, NJ: Prentice Hall.

SMITH, A. (1776) *An Inquiry into the Nature and Causes of the Wealth of Nations* (Cannan Edition) New York: Modern Library, 1937.

SMITH, R.K. (1987) 'Explaining the Non–Proliferation Regime: Anomalies for Contemporary International Relations Theory', *International Organization* 41, 2 (Spring): 251–81.

SNIDAL, D. (1979) 'Public Goods, Property Rights, and Political Organizations', *International Studies Quarterly* 23, 4 (December): 532–66.

—— (1985) 'The Limits of Hegemonic Stability Theory', *International Organization* 39, 4 (Autumn): 579–614.

SPROUT, H. and M. SPROUT (1971) *Toward a Politics of Planet Earth*, Princeton, NJ: Van Nostrand.

STEIN, A.A. (1982) 'Coordination and Collaboration: Regimes in an Anarchic World', *International Organization* 36, 2 (Spring): 299–324.

STERLING, R.W. (1970) *Macropolitics: International Relations in a Global Society*, New York: Alfred A. Knopf.

STRANGE, S. (1976) *International Monetary Relations*, New York: Oxford University Press.

—— (1979) 'The Management of Surplus Capacity: Or How Does Theory Stand Up to Protectionism 1970s Style?', *International Organization*, 33, 3 (Summer): 303–34.

—— (1982a) 'Cave! Hic Dragones', *International Organization* 36, 2 (Spring): 479–96.

—— (1982b) 'Still an Extraordinary Power', in R. Lombra and W. Witte (eds) *The Political Economy of International and Domestic Monetary Relations*, pp. 73–103, Ames: Iowa State University.

—— (1984) 'The Hegemon's Dilemma', *International Organization* 38, 2 (Spring): 287–304.

—— (1985) 'Protectionism and World Politics', *International Organization* 39, 2 (Spring): 233–59.

—— (1987) 'The Persistent Myth of Lost Hegemony', *International Organization* 41, 4 (Autumn): 551–75.

STRANGE, S. and R. TOOZE (eds) (1981) *The International Politics of Surplus Capacity*, London: Butterworth.

TAYLOR, P. (1968) 'The Functionalist Approach to the Problem of International Order: A Defense', *Political Studies* 16: 393–409.

—— (1975) 'Functionalism: The Theory of David Mitrany', in A.J.R. Groom and P. Taylor (eds) *Functionalism: Theory and Practice in International Relations*, pp. 236–51, London: University of London Press.

THORNTON, R. (1970) *International Airlines and Politics*, Ann Arbor: Bureau of Business Research, Graduate School of Business Administration, University of Michigan.

—— (1971) 'Governments and Airlines', *International Organization* (Summer).

TSOUKALIS, L. and A. da SILVA FERREIRA (1980) 'Management of Industrial Surplus Capacity in the European Community', *International Organization* 34, 3 (Summer): 355–76.

VAN DEN NOORT, P.C. (1983) 'The Common Agricultural Policy Key to European Economic Integration', in P. Coffey (ed.) *Main Economic Policy Areas in the EEC*, The Hague: Martinus Nijhoff.

VERNON, R. (1971) *Sovereignty at Bay: The Multinational Spread of US Enterprises*, New York: Basic Books.

—— (1974) 'Enterprise and Government in Western Europe', in R. Vernon (ed.) *Big Business and the State: Changing Relations in Western Europe*, Cambridge, MA: Harvard University Press.

VOGEL D. (1986) *National Styles of Regulation: Environmental Policy in Great Britain and the United States*, Ithaca, NY: Cornell University Press.

WALLERSTEIN, I. (1976) *The Modern World-System: Capitalist Agriculture and the Origins of the European World Economy in the Sixteenth Century*, New York: Academic Press.

WALTZ, K. (1979) *Theory of International Politics*, Reading, MA: Addison-Wesley.

YOUNG, O.R. (ed.) (1975) *Bargaining: Formal Theories of Negotiation*, Chicago, IL: University of Illinois Press.

—— (1979) *Compliance and Public Authority*, Washington, DC: Resources for the Future.

—— (1980) 'International Regimes: Problems of Concept Formation', *World Politics* 32 (April): 331–56.

—— (1982) 'Regime Dynamics: The Rise and Fall of International Regimes', *International Organization* 36, 2 (Spring): 277–98.

—— (1986) 'International Regimes: Toward a New Theory of Institutions', *World Politics* 39, 1 (October): 104–22.

ZACHER, M.W. (1987) 'Trade Gaps, Analytical Gaps: Regime Analysis and International Commodity Trade Regulation', *International Organization* 41, 2 (Spring): 173–202.

ZARTMAN, W. (1975) 'Negotiations: Theory and Reality', *Journal of International Affairs* 9, 1 (Spring): 69–77.

—— (1976) *The 50% Solution*, Garden City, NY: Anchor Press, Doubleday.

ZYSMAN, J. and L. TYSON (eds) (1983) *American Industry in International Competition: Government Politics and Corporate Strategies*, Ithaca, NY: Cornell University Press.

Personal Interviews

BENDELL, Elaine – Public Relations Manager, McDonnell-Douglas Corporation, 3 January 1983.

BRACKEN, Nicole – Senior Market Analyst, Allied/Bendix Aerospace, 26 February 1986.

LEDERER, Jerry – President Emeritus, Flight Safety Foundation, 10 April 1985.

KOPLIN, Klaus – Chief, Engineering Division, West Germany's Civil Aviation Authority, 11 July 1987.

STEGGARDA, Paul – Business Planning, Sperry Commercial Flight Systems, Division Honeywell, 6 October 1987.

WILY, Richard – Director of Airline Market Analysis, The Boeing Commercial Airplane Company, 27 February 1986.

Personal Correspondence

BACHER, Thomas (1984; 1987) Director of International Business, The Boeing Commercial Airplane Company (17 March; 25 October).

BOUMA, W.J. (1983) Managing Director of the Aeronautical Inspection Directorate of The Netherlands' Department of Civil Aviation (30 March).

BOWIE, David (1982) Department of Commerce (5, 8 April).

CHAPLIN, J.C. (1983) Director-General of Airworthiness, Civil Aviation Authority of the United Kingdom.

CLIFFORD, Douglas (1981) Chief Engineer of Flight Management Systems, The Boeing Commercial Airplane Company (15 June).

DeLAUER, Joseph (1980) Chief, Flight Standards Division, Western Region, Federal Aviation Administration (23 December).

EDMONDS, Martin (1978; 1980) Professor, Department of Politics, University of Lancaster (21 August; 13 January).

ELLE, B.J. (1984) Chief, Economics and Statistics Branch, International Civil Aviation Organisation (6 June).

FERRER, James (1982) Director of the Office of Aviation, Department of State (12 April).

GANSLE, James (1982) Department of Transportation, International Aviation Division (10 May).

GRAVATT, Joan (1982) Department of State, Foreign Service Officer in the Aviation Programs and Policy Division (24 April).

GUTH, Herbert (1982) Air Transport Specialist, National Academy of Sciences, Transportation Research Board (11 May).

HIGHAM, Robin (1981) Editor of *Aerospace Historian* and Professor of History at Kansas State University (12 April and 7 August).

JENNY, Larry (1987) Aviation Specialist, National Academy of Sciences, Transportation Research Board (19 November).

KENNEDY, A.P. (1981; 1982; 1983) Secretary, Joint Airworthiness Requirements (16 July; 31 March and 13 May; 11 February).

KING, James (1981) Chairman, National Transportation Safety Board.

LEDERER, Jerry (1982; 1987; 1988) President Emeritus, Flight Safety Foundation (6 May; 17 November; 12 February).

LOOS, James (1982) Chief, International Liaison and Policy Division, Federal Aviation Administration (3 May).

LOPATKIEWICZ, Ted (1982) Civil Aeronautics Board, Public Affairs Office (4 April and 10 May).

MADAYAG, A.F. (1981) Flight Standards Division, Western Region, Federal Aviation Administration (3 June).

MATT, John (1981) Acting Chief, Operations Liaison Branch, Office of International Aviation, Federal Aviation Administration (9 June).

MOUDEN, L. Homer (1981; 1982) Vice–President Technical Affairs, Flight Safety Foundation (29 January; 27 September).

NORLING, Alfred (1982) Aviation Specialist, Kiddre Peabody & Co. (23–6 May).

PLUMMER, Norman (1980; 1982) Deputy Director, Office of International Aviation, Federal Aviation Administration (23 December; 16 April).

POUR, George (1982) Chief, Aircraft Manufacturing Division, Office of Airworthiness, Federal Aviation Administration (11 June).

RUSSELL, Robert (1982) Staff Director, International Economic Policy Subcommittee, Committee on Foreign Relations, United States Senate (23 June).

SEELEY, Ken (1982) Director of Education, Professional Aviation Maintenance Association (28 April).

SHAW, R.R. (1981) Assistant Director General–Technical, International Air Transport Association (9 January).

THAYER, Frederick (1987) Professor, Graduate School of Public and International Affairs, University of Pittsburgh (22 October, 19 November).

Index